Active Pharmaceutical Ingredients

Ingredients

Development, Manufacturing, and Regulation

DRUGS AND THE PHARMACEUTICAL SCIENCES

Executive Editor

James Swarbrick

PharmaceuTech, Inc.
Pinehurst, North Carolina

Advisory Board

DRUGS AND THE PHARMACEUTICAL SCIENCES

A Series of Textbooks and Monographs

Active Pharmaceutical Ingredients

Development, Manufacturing, and Regulation

edited by

Stanley H. Nusim
S. H. Nusim Associates, Inc.
Aventura, Florida, U.S.A.

Taylor & Francis
Taylor & Francis Group

Boca Raton   London   New York   Singapore

Published in 2005 by
Taylor & Francis Group
6000 Broken Sound Parkway NW, Suite 300
Boca Raton, FL 33487-2742

International Standard Book Number-10: 0-8247-0293-X (Hardcover)
International Standard Book Number-13: 978-0-8247-0293-9 (Hardcover)

Library of Congress Cataloging-in-Publication Data

Catalog record is available from the Library of Congress

Taylor & Francis Group
is the Academic Division of T&F Informa plc.

Visit the Taylor & Francis Web site at
http://www.taylorandfrancis.com

Preface

Active pharmaceutical ingredients known today as "APIs" are organic chemicals, generally synthetic, that are the subject of this book. These ingredients are chemicals that will be used in a final pharamaceutical dosage form. The manufacturing of these chemicals is a subsection of fine chemical manufacturing. This subsection of the chemical industry has undergone very significant changes in much the same manner, but perhaps trailing, the pharmaceutical industry that manufactured the final dosage form.

The "pharmaceutical industry" at the turn of the 20th century was essentially the local pharmacy (or "chemist" as it was also known). The "bulk pharmaceutical chemical industry" at that time was merely a provider of all those laboratory chemicals, including solvents and excipients as well as APIs needed by the local pharmacist to compound the prescribing doctor's formulation.

Over this past century, as with many industries, enormous changes have occurred in the pharmaceutical industry, causing equally significant changes for API suppliers. It is these changes, many of which have accelerated in recent decades, that suggested the need for a definitive reference for this manufacturing activity.

At one time following routine chemical manufacturing practices would have been sufficient; however, this is no longer the case. Not only has there been a significant shift in the government regulations that control the redefined "quality" of the product, but a very intensive look at the development of the process to be used as well as the manufacturing activities required to make the API.

This focus is to ensure that the API is produced in an environment that ensures it is free of contamination that may be introduced from inherent process impurities but also from the manufacturing environment itself. The latter is controlled by the so-called "cGMPs" (current Good Manufacturing Practices), while the former by the nature of the chemical process and the level of quality assurance that the process provides; hence, a focus on the process development is essential.

It is the intent of this volume to focus on the three overall activities that bring an API to market; the development of the chemical process, the manufacturing activity utilizing that process, and the governmental regulations that control the approval of the product so that it may be commercially marketed. This book brings together information into a single source that will allow those in the field to be sure they are up to date. In addition, it will provide to those organizations that are planning to enter this field, the basic information needed to think through, understand, and effectively plan bulk manufacturing of an API.

The rapidly changing environment that has occurred in the past decades shows no signs of easing; thus, this volume will be a starting point. Ongoing continuing attention to all aspects of these issues is an absolute necessity to ensure that manufactured APIs will meet the newest standards in an environment that has seen many changes in the market itself as well as its regulation, product mix, and volume.

This text covers those three activities of development, manufacturing, and regulation in its broadest sense. This will include discussions on the process development cycle, introduction of the process into factory design engineering, regulatory matters that include the regulatory approval process, quality control/assurance, and validation as well as the standard plant manufacturing operation activities including materials management and planning and maintenance. In addition, it will discuss other plant operational issues including safety and environmental issues that are part of any chemical manufacturing operation.

I have chosen to exclude fermentation and other biological processes from this book although products from those processes continue to be an increasingly important source of pharmaceutical actives in today's world. This decision was made because the chemical routes remain the largest source of actives to the pharmaceutical industry. Actives supplied by biological processes are no less important than chemically generated actives but are sufficiently different to be worthy of their own volume.

I wish to express my thanks to the publisher for its invitation to assemble this book and particularly to Sandra Beberman for bearing with me in the very long and tedious development process for the book. Her advice and encouragement throughout this process was a primary driving force to ensure its completion.

Stanley H. Nusim

Contents

5. Regulatory Affairs *167*
John Curran

6. Validation *203*
James Agalloco and Phil Desantis

7. Quality Assurance and Control *235*
Michael C. VanderZwan

Contributors

James Agalloco Agalloco & Associates, Inc., Belle Meade, New Jersey, U.S.A.

Eugene Bobrow Merck & Co., Inc., Whitehouse Station, New Jersey, U.S.A.

Victor Catalano Purchasing Group Inc. (PGI), Nutley, New Jersey, U.S.A.

John Curran Merck & Co., Inc., Whitehouse Station, New Jersey, U.S.A.

Phil Desantis Schering-Plough Corp., Kenilworth, New Jersey, U.S.A.

Steven Mongiardo Merck & Co., Inc., Whitehouse Station, New Jersey, U.S.A.

Stanley H. Nusim S. H. Nusim Associates Inc., Aventura, Florida, U.S.A.

Raymond J. Oliverson HSB Reliability Technologies, Kingwood, Texas, U.S.A.

Carlos B. Rosas   Rutgers University, New Brunswick, New Jersey, U.S.A.

Michael C. VanderZwan   Pharmaceutical Technical, Roche Pharmaceuticals, Basel, Switzerland

1

Introduction

STANLEY H. NUSIM

S. H. Nusim Associates Inc., Aventura, Florida, U.S.A.

Pharmachemical manufacturing is that branch of fine chemical manufacturing that is directed to the manufacture of chemicals whose ultimate use will be in a final pharmaceutical dosage form, referred to as the active pharmaceutical ingredient ("API"). This industry segment has undergone very significant changes in much the same manner, but trailing, the pharmaceutical industry itself, from the time it emerged early in the 20th century.

Thus, we must examine what has happened in the pharmaceutical industry over this period, in order to understand the implications for API manufacturing. This will lead us to the present time and to the goal of this book.

It is our objective to provide a reference book that speaks to those issues that need to be addressed in order to assure that an existing or proposed pharmachemical operation will meet its objective of supplying an API to meet a medical/market need efficiently and effectively.

The changes are themselves a result of major changes that have occurred both directly and indirectly, in and on, the industry. These changes include company consolidations, both backward and forward integration, the increased and changed role of quality, the significant intensification of regulatory bodies worldwide, the impact of the greatly increased potency of APIs, thereby reducing pharmachemical requirements and the broadening of the market worldwide.

These ideas will be discussed briefly here and touched on in depth in the subsequent chapters.

I. CONSOLIDATION AND INTEGRATION

The "pharmaceutical industry" at the turn of the 20th century was essentially the local pharmacy (or chemist as it was also known outside of the United States). The objective of the pharmachemical supplier to the local industry, at that time, was a provider of all the chemicals, including APIs, as needed by the pharmacist to compound the prescribing doctor's prescription.

Thus, the great pharmaceutical titans of today, such as Merck, were a fine chemical manufacturer providing a full variety of basic laboratory chemicals and solvents as well as the actives of the day, in order to meet all of the formulating needs of the pharmacist. This activity was common in those early days, as well, to Pfizer, Bayer, and Sterling, among others.

The forward integration of these companies into providing the finished dosage form had by the middle of this past century become the standard rather than the exception, as the medical community shifted to writing prescriptions for the local pharmacist to fill prescribing finished dosage forms rather than their own formulations. This practice continues today, as the determination of efficacy and safety of formulated product has grown to very significant proportions.

II. QUALITY

An overriding driving force in this direction, although it may never have been originally intended, has been the shift of governmental control that has been exercised by the U.S. Food & Drug Administration ("FDA"). A brief discussion of that change is now in order.

The initial purpose of the first Pure Food, Drug & Cosmetics Act (Act) that was passed by Congress in the first decade of the last century was one of safety. It began by the regulation of those items of commerce that had the potential of poisoning the individual who used it if the product was contaminated. It is for this reason that the Act covered those three specific items, all lumped together although each being used for very different purposes.

The initial focus, at that time, for drugs as well as the other two types of ingested or topically applied products, was lack of contamination as determined by quality sampling and testing. In addition, and extrapolating that issue to new proposed pharmaceuticals, the key data required was the toxicity data and its ratio to the proposed dose level, the "therapeutic index." However, no data or judgment on efficacy was required for its proposed use. Its medical purpose and its ultimate use remained in the hands of the physician and the sponsoring company that promoted it.

In the middle 1950s, this changed dramatically when the Act was amended significantly. The change, driven by congressional hearings and the "Thalidomide affair,"[a] now required not only more significant safety data, beyond simple toxicity, but also, more significantly, scientific proof

---

[a]Thalidomide was an antinausea drug that at that time had been approved in Europe and was before the FDA for approval in the United States. Pregnant women who are normally prone to nausea became an instant market for the new drug. However, very serious birth defects (missing limbs) were experienced in babies born to many of those women who had taken the drug. This precipitated a worldwide reaction to review the new drug approval process. Needless to say that the drug was not approved in the United States at that time. (In recent years, it has been approved for limited special use in leprosy.)

of efficacy. This now placed a new burden on the sponsoring company to provide unequivocal proof, to the government's satisfaction, that the addition of a new chemical entity at the dose level recommended was worthwhile to the public. The shift was due to the recognition that replacing a tried and true medication, that was widely used and its side effects well defined with a new compound with only limited experience in man, was in itself an unknown risk and, therefore, must be shown to be worth the risk.

This greatly increased the cost and the risk associated with the discovery and introduction of new chemical entities. This change was absorbed by the industry and set the stage for the next major shift in policy that came in the middle 1970s. This was the establishment of current good manufacturing practices (cGMPs) for the manufacture of pharmaceutical actives as well as the finished pharmaceutical products.

This was the next step in the focus of the FDA on the safety of the product. Up until this point, contamination (or lack thereof) was defined by the presence (or absence) of foreign impurities not specified in the analytical protocol for the product. This was the case for either the pharmaceutical product or the API that went into the finished product. Although this could be a definitive test for a uniformly distributed contaminant, it would not necessarily find random contamination that occurred in processing or extraneous matter that could enter the system from dirty facilities or poor operating practices.

Finished goods testing, today, as it was at that time always depended upon the assumption of uniformity of product. It was this presumption that permitted the approval and release of a product based on the testing of 100 g of a 100 kg pharmachemical batch or 30 tablets of a lot of 500,000 tablets.

The concept of current "cGMPs" and quality assurance became the dominant theme, thereby pushing the analytical testing (quality control) into the background.

In principle, one now had to show, in order to have a product free of contamination, that the manufacturer produced the product in contaminant-free equipment in a clean facility, within equipment designed and tested to show consistent and reproducible product by people thoroughly trained and with full knowledge of the process. Thus, in the United States, this greatly shifted the emphasis to a more rigorous standard of "quality".

The most recent change implemented is the requirement of formal "validation" of facilities, equipment and the process itself. This is the "proof" that the process and the facility can produce quality product on a consistent basis.

In a similar fashion, one can see the extension of the tighter regulations as they apply in the United States, to Japan and Western Europe. Through the EU, they have implemented similar standards for the very same reason in Europe; additionally, many of the "third world" nations have already implemented their own GMP initiatives reemphasizing the growing uniformity in such requirements throughout the world.

All these factors are discussed more thoroughly in the appropriate chapters within this book.

III. POTENCY

A subtle change that has emerged in the methods of discovering and developing new drugs in the past decades has had significant impact on the pharmachemical industry.

In the early days, the key to drug discovery often was screening programs, where laboratory-screening models were used to test new chemical entities for efficacy against specific disease candidates. Those that were effective, however, often found much of their potency diminished as the active, generally formulated into a pill, was attacked by normal body chemistry as it passed through the digestive system on its way to be absorbed into the blood and transported to the disease site. Thus, only a fraction of the orally ingested drug reached the drug target area. As a result, dose regimens for most oral drugs were 100–500 mg.

These dosing levels generated needs for significant quantities of actives in some cases into the millions of kilograms annually. (Five billion tablets at 200 mg dose require 1 million kilograms of active.) This resulted in dedicated plants for each drug active, particularly since the active was generally a complex organic molecule requiring many chemical steps to synthesize.

However, with the advent of the focus on biochemistry and the new sophistication gained in understanding the chemistry and biology of the body, today's drugs are designed so as to be more potent. In addition, they can be chemically protected to limit the destruction of the drug as it passes through the body on its way to the target site. Thus, normal dosing of today's "designer" drugs are 5–20 mg, 10 fold less than in the past. This reduces the API need for "blockbuster" drugs by an order of magnitude (5 billion tablets at 10 mg dose requires 50,000 kg of API). This also suggests that those lesser volume products would require very small quantities of API making dedicated facilities for them very uneconomical.

These factors have refocused API manufacturing from facilities dedicated to a single API product to multiproduct manufacturing facilities. The added costs of a facility due to the more rigorous cGMPs that now apply favor these kinds of facilities, where the cost can be shared by many rather than a single product.

This adds a very critical aspect to the operation because the issues of equipment clean out and turnaround, particularly as the issue of cleanliness to ensure that cross contamination does not occur must be addressed.

IV. COMPUTER CONTROL AND AUTOMATION

This industry, like nearly all others, has seen the positive impact of the introduction of computers and automation in the manufacturing facilities. The first impact was in the automatic control systems that are used to maintain accurate and reproducible operating conditions for reaction and isolation systems. This was extended into the integration of multiple operations under computer control often eliminating or at least minimizing people intervention.

This itself caused some concerns for the FDA, which, in the past, depended upon manual documentation by operators of batch procedures written and issued by people and people observing and recording all data. This was transformed to computer-recorded data and operating instructions being maintained in computer files. This generated an entire series of new issues that had to be dealt with by both the operation and the FDA. First was security to be sure that the automated instructions are safe from improper and unauthorized changes to the issue of signatures, often electronic signatures, a new concept that has become very common.

V. SUMMARY

The changes referred to above, and the changes that are to occur, without doubt, in the future, drive the need to understand where we are today and where we are going in the future. We have chosen to address the various segments and activities of a pharmachemical plant by having a focused discussion on each in the subsequent chapters.

Again, I repeat a statement from the Preface. Each and every topic covered in this volume has changed from the past and will continue to change in the future; therefore, the reader is being presented with a "starting point" from which he or she must continue to follow the progress of in order to keep current.

2

Process Development

CARLOS B. ROSAS

Rutgers University, New Brunswick, New Jersey, U.S.A.

I. INTRODUCTION

The purposes of this chapter are few and rather ambitious. First, to provide a sound perspective of bulk drug process work to the uninitiated and the relatively new practitioner, hopefully without prejudice to the benefit that the approach herein might afford to an experienced but still restless practitioner. All work in a forest that is dense and rich in its variety; it should be regarded from a vantage now and then, and it is from such a deliberately selected vantage that the chapter unfolds.

Then there is the promotion of the power that the purposeful convergence of chemistry, microbiology, and chemical/

9

biochemical engineering can bring to bear on the increasingly difficult task at hand: *the timely conception, development, and reduction to practice at scale of a sound process for the manufacture of a bulk drug.* In the 2000s, timely is shorthand for swift, sound encompasses safety to the environment and to people as well as amenable to various regulatory approvals, and reduction to practice at scale means that the resulting process can be used for reliable manufacture in whatever context might be first required.

> Chemistry, in the context at hand, is the aggregate of synthetic, analytical, and physical chemistry fields within what may be called the drug *process* chemistry discipline at large. The latter, while practiced for decades, has truly come into being in the 1990s, spurred mostly by the greater ascendance of the pharmaceutical industry among chemistry practitioners and by the enhanced role of the bulk drug process in the outcome of drug development. Whereas toxicology or clinical results were the exclusive causes for the demise of drug candidates, the greater difficulty in making today's more complex structures in today's regulatory milieu has for some time raised the profile of their bulk process development task as a factor in the overall outcome (1).

> Although first manufacture of the bulk drug is the paramount objective of the technology transfer to manufacturing, the process body of knowledge should be sturdy and complete enough to support expanded manufacture for product growth, as well as provide at least a clear sense of direction for process improvements or second-generation processing.

The above definitions conveniently describe a complex task, to which considerable skills need to be applied with due deliberation and under constant managerial attention. Indeed, successful bulk drug process development, as just defined, requires that sufficient interdisciplinary and operational resources be brought together in a cohesive manner, not unlike that required by a critical mass in nuclear fission. Most often, having the resources is not enough, and their cohesiveness makes a significant difference in the degree of success, sometimes making the ultimate difference: *having or not having a new drug available when needed.*

Another sought perspective applies to the integration of the bulk drug process development task with the simultaneous drug development program at large: toxicology, dosage form development, clinical development, and the assembly of the regulatory submissions. The latter, leading to the desired regulatory approvals as the culmination of the overall effort, has in recent years become increasingly dependent on the scope and execution of the process work for the bulk drug, which in some of its aspects has now become fastidious and greatly increased the burdens of the bulk process development task.

As the last objective, the methods of bulk drug process development will be weaved discreetly, if not seamlessly, throughout the chapter: (a) the principal issues that shape the methods, (b) the most trenchant choices confronting the process development team, and (c) some selected heuristics (i.e., empirical rules that, although lacking proof, are useful often enough) distilled from the author's experience.

> As a distinct and credible literature of process development for bulk drugs and fine chemicals has come into being and grows, statements of applicable empirical wisdom are appearing with a modicum of organization (2–6) and the field should one day become amenable to independent study (it is not currently taught formally anywhere). In addition, a journal focused on the field has been published since 1997 as a joint venture of the American Chemical Society and the Royal Chemical Society (7). Alas, the engineering scale-up of synthetic bulk drug processes is still badly understated, as most contributors to the new body of literature are synthetic chemists. For compounds derived from biosynthesis, however, there is a large body of biochemical engineering literature that deals in depth with the scale-up of the biosyntheses and the subsequent "downstream processing" technologies (e.g., Refs. 8–10).

The application of the fruits of bulk drug process development to process design, technology transfer, and first manufacture will be addressed in the companion chapter 3, as those activities are carried out in a distinct context that overlaps with the R&D activities. Such planes of contact will, of course, be identified in this chapter and their discussion confined to the minimum needed herein.

About the scope of the chapter, it is ambitious in its aim to support the above objectives, yet modest in its depth of descriptive material, since doing justice to the latter would require a much larger volume. Instead, the author has chosen to address the fundamentals along the said objectives, while keeping the descriptive technical material spare and aimed at selected targets of the bulk drug process development task: e.g., seeking thermochemical safety, scaling-up, achieving the desired physicochemical attributes of the bulk drug, capturing and applying the process know-how.

As of this writing in 2004, the process development milieu of the bulk drug industry is quite varied—from the large drug company in which all the skills are represented to the small virtual firm that contracts out the work, as well as firms that do selected process development tasks as part of their attempt to secure the eventual manufacturing business from the owner of the drug candidate. The author has not attempted to deal separately with these different environments lest the exposition of the target fundamentals get obscured by the specifics of each case. Instead, the bulk drug process development task is discussed within the continuum of a large drug company and commentary that applies to other contexts has been inserted, hopefully in a sparing and incisive manner.

The reader should be alerted to an additional choice of the author. Although the increased regulatory expectations have deeply transformed the process development task, the paramount stance for the practitioner remains intact: *know and understand your process, reduce it to practice soundly, and operate it in a disciplined manner.* Accordingly, this and its companion chapter, aimed at the fundamentals, avoid the spectrum of the current good manufacturing practices subject (or cGMP), which seems to have soaked so much of the energy of process practitioners throughout the bulk drug industry. However, the issues associated with the assembly of regulatory submissions (IND, NDA and the like) and with the expectations of the subsequent approval process will be discussed as required to meet the objectives of the chapters.

Finally, the diligent reader of these two chapters, armed with the perspectives provided herein, should find that continued study of the literature can be quite fruitful. To assist in that task, a selection of references is included, most of which are cited throughout the text, with the rest cited separately as suitable reading for the studious.

II. THE BULK DRUG PROCESS AS PART OF THE DRUG DEVELOPMENT PROGRAM

A. The Chemical Process of a Bulk Drug

In the context of this chapter, a *bulk drug* or a *bulk drug substance* is a material—a single chemical compound with the desired biological activity—obtained in bulk form and destined for the preparation of dosage forms. The latter, when administered in a prescribed manner to the target patient, animal or plant, delivers the drug so as to elicit a desired physiological response and, in due course, the intended therapeutic or protective result. More recently, terms such as *active pharmaceutical ingredient (API)* or *bulk pharmaceutical chemical (BPC)* seem to have overtaken the usage, seemingly as the result of their adoption by regulators in the United States. Herein we will use the original term *bulk drug* (or *bulk*), as it most aptly describes the material—a drug that is obtained and characterized in bulk form. However, we will confine our scope to those compounds commonly known as *chemical entities*—drugs of relatively small molecular weight that can be characterized well by current methods of chemical and physicochemical analysis. In doing so, we are excluding those macromolecules, substances, and preparations of biosynthetic origin that are collectively known as *biologicals*. The processing methods used in biologicals, albeit based on the same fundamentals, are significantly different from those applied to chemical entities, and their process development, registration, and manufacture also take place in a rather different environment. In addition, organic compounds categorized as nutritionals and fine chemicals at large are not within this scope, their processing similarities with bulk drugs notwithstanding.

Bulk drugs are obtained through three chemical processing routes:

    a. Extraction, recovery, and purification of the drug from biomasses of natural origin or from fermentation (Fig. 1): (1) paclitaxel is extracted from various *Taxus* plants; (2) lovastatin is biosynthesized in the fermentation of nutrients by *Aspergillus terreus*.

    b. Semisynthesis, in which a precursor compound from a natural source or a fermentation is converted to the target drug by synthetic chemical modification: (1) fermentation penicillin G is converted to 6-aminopenicillanic acid, which in turn is reacted with an acyl chloride to afford ampicillin; (2) natural morphine is methylated to codeine (Fig. 2).

Both routes to bulk drugs take advantage of the diversity and richness of molecular structures found in natural sources, where many important biological activities are also found.

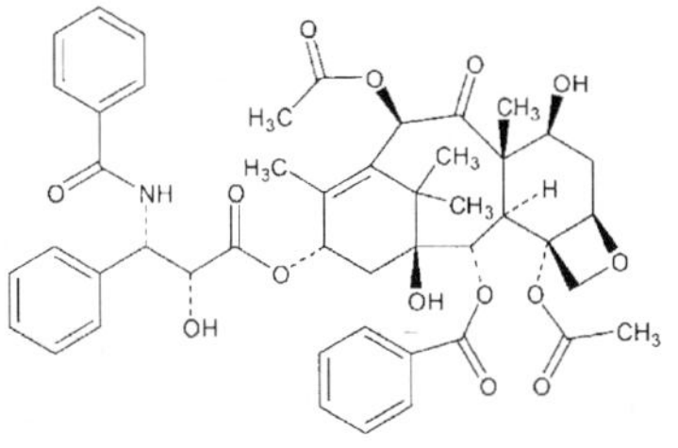

Paclitaxel – Found in *Taxus* plants.

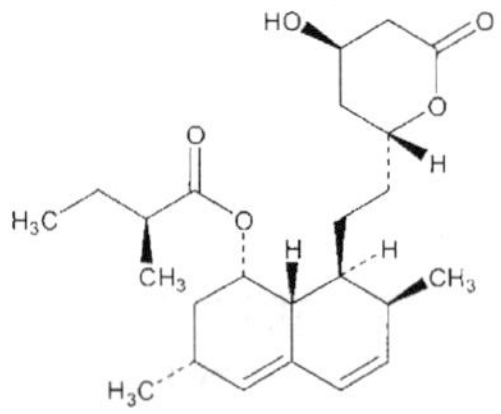

Lovastatin - From certain microbial cultures.

Figure 1 Bulk drugs from natural sources: Paclitaxel (antileukemic and antitumor) and lovastatin (inhibitor of cholesterol biosynthesis) are examples of the diverse and complex structures made by plant and microbial cell biosyntheses, respectively. In most instances of such compounds having desirable biological activities, their structural and chiral complexities make chemical synthesis not competitive with isolation from biosynthesis.

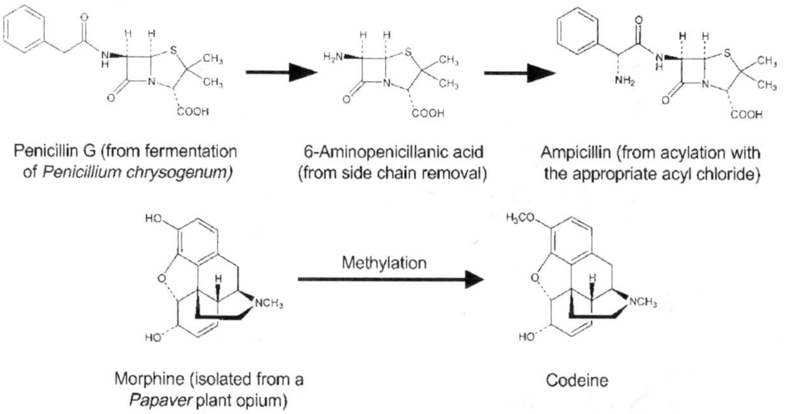

Penicillin G (from fermentation of *Penicillium chrysogenum*)

6-Aminopenicillanic acid (from side chain removal)

Ampicillin (from acylation with the appropriate acyl chloride)

Methylation

Morphine (isolated from a *Papaver* plant opium)

Codeine

Figure 2 Semisynthetic bulk drugs: Ampicillin (antibacterial) from penicillin G. Modifications of biosynthetic structures are often created to improve the in vivo attributes of the original compound, utilizing the biosynthesis product as the starting material containing most, if not all, of the structural complexity that provides the basic biological activity. Similarly, codeine (analgesic), although found in opium from *Papaver* plants, is most economically made by methylation of morphine, which is more efficiently isolated from opium.

    c.   Total synthesis from simple starting materials or less simple intermediate compounds (Fig. 3): (1) fosfomycin from commodity chemicals, (2) lobetalol from 5-bromoacetyl salicylamide.

    In either total synthesis or semisynthesis processing, sometimes a desired synthetic transformation is best done by an enzyme. Such synthesis step, whether using a preparation of the enzyme or the host microorganism, will be considered a chemical synthesis step (a *biotransformation* or a *biocatalytic step*) and not a fermentation for biosynthesis.

Whichever of these routes is used to obtain a bulk drug constitutes the *chemical process*. Further processing of the bulk drug to obtain the dosage form constitutes the *pharmaceutical process*. This distinction is depicted in Fig. 4, where simple

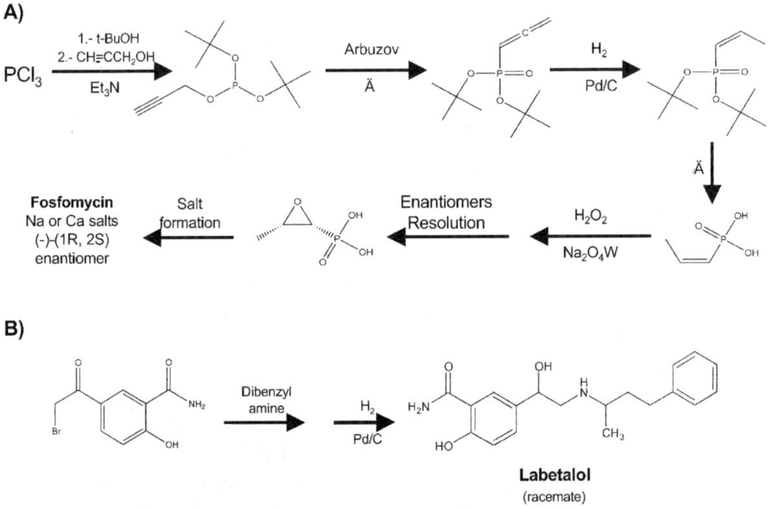

A)

B)

Labetalol
(racemate)

Figure 3   Drugs by total synthesis: Fosfomycin (antibacterial) is a good example of the manufacture of a bulk drug by total synthesis from basic chemicals, albeit the compound is of biosynthesis origin. Alternatively, and more frequently, the manufacturing process is simplified by tapping on commercially available compounds of greater structural complexity (*intermediates*), such as 5-bromoacetyl salicylamide as the starting material for lobetalol (antihypertensive). Even if the intermediate is custom made by others, the process development and manufacturing task for the drug developer is greatly simplified relative to the use of basic or building block chemicals.

graphical means are used in an attempt to differentiate the *bulk* character of the product of the chemical process, in contrast to the *discrete* character of the product of the pharmaceutical (or *dosage form* or *secondary manufacturing*) process. The distinction also reflects their very different technology, manufacturing, and regulatory environments.

In the current pharmaceutical parlance, the term API is used most often as descriptive of the biological activity contribution. Herein, however, the term *bulk drug* is used instead as descriptive of the physical and chemical character of the subject material, with its biological activity taken as obvious. Indeed, the conventional term for

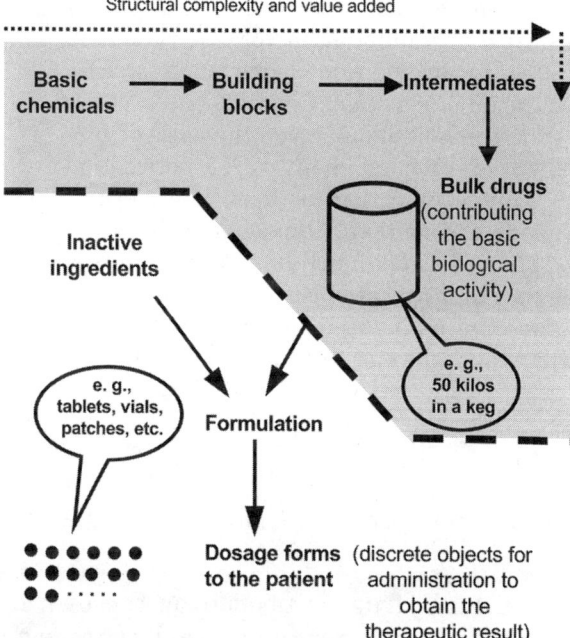

Figure 4 The domains of chemical (bulk drug) and pharmaceutical (dosage form) processing, with the chemical processing domain defined by the shaded area of the diagram.

the other ingredients added to formulate the dosage form is still *inactive pharmaceutical ingredients*.

As we proceed, unavoidably some other terms will be used that may not be familiar to all readers. Accordingly, an effort will be made to define such terms at the point of first use, as well as to use them sparsely. For example, *unit operations* are those methods that can be found repeatedly used in chemical processing and that have a common phenomena root, their many variations notwithstanding—filtration to separate solids from an accompanying liquid, distillation to separate volatile components from a mixture, or milling to reduce the particle size of particulate solids. The organization of chemical processing on the basis of such unit operations was crucial to the development of organic chemical technology, which was originally arranged on the chemistry basis of *unit processes*, such as nitration, sulfonation, or esterification. Whereas the latter organized knowledge on a strictly descriptive basis, the unit operations approach made possible the study of processing phenomena on the basis of generalized principles from physics, chemistry, kinetics, and thermodynamics, which could then be used to undergird methods applicable in the context of any chemical process and over a wide range of scale and circumstances. Hence, the keystone role that unit operations played in the advent of chemical engineering as a discipline, with a practice quite distinct from that of the earlier industrial chemistry.

B. A Perspective

Process development of a bulk drug consists of three distinct tasks:

   a. Preparation of bulk drug as needed by the overall development effort—*the preparative task*. The scope of this task varies over a wide range, as shown in Table 1.
   b. Definition and achievement of the desired physico-chemical attributes of the bulk drug as needed by the dosage form development—*the bulk drug definition task*.
   c. Acquisition and organization of a body of knowledge that describes a sound process for regulatory submissions and technology transfer to first manufacture at scale—*the body of knowledge task*.

However, these tasks cannot be directed to successful and timely completion unless viewed and managed as a veritable trinity, their differing demands and instantaneous urgencies notwithstanding. Drug development is a fast paced and difficult enterprise; it presents frequent junctures at which the need to focus on the most compelling task needs to be artfully balanced with other needs lest the aggregate task be compromised—all three tasks need to be completed

Table 1 Bulk Drug Demands of the Various Drug Development Phases

Preclinical phase—Initial toxicology, probes on drug bioavailability, data gathering for the IND, additional animal studies, etc.	Supplies to be delivered over 2–6 months	Total ~5–50 kg
Phase I—Use in humans (20–80 mostly healthy subjects) for pharmacokinetic, pharmacological, routes of administration, dose ranging, and tolerance studies. Continuing toxicology and dosage form development. All aimed at the design of Phase II/III studies and defining the target dosage forms	Supplies to be delivered over 6–12 months	Total ~20–100 kg
Phase II/III—Increasingly large number of patients (up to thousands) in studies for therapeutic effectiveness (initial and confirmatory), dose and regimen determination, evaluation of target populations for safety and efficacy, support of desired claims, market-specific and dosage form-specific studies, etc. Continuing toxicology and dosage form development, stability studies. All aimed at the assembly of the dossier	Supplies to be delivered over 18–48 months	Total ~300 to > 2000 kg.
Phase IV—Postapproval studies for optimization of drug use, pharmacoeconomic data, morbidity and mortality data, head-to-head and concomitant drug uses, etc.	These studies are generally supplied from bulk drug made in the manufacturing operation.	

Notes:
1. IND (investigational new drug) is the submission requesting USFDA's exemption from drug shipping in interstate commerce, thus signaling the intent to initiate study in humans (or target species if a veterinary drug). Dossier is a term often used to describe the total body of knowledge on the drug candidate, from which individual submissions are assembled for filing with the various agencies; e.g., the new drug application (NDA) to the USFDA.
2. The ranges of bulk drug totals reflect the wide differences among drug candidates and their programs. Issues such as drug potency and dosage regimens, low animal toxicity, length of treatment to the clinical endpoint, relative difficulty of dosage form development, number of dosage forms developed and scope of the clinical studies are the principal factors determining the demands for bulk drug. Obviously, relatively infrequent extremes exist on both ends: from a low end for drugs such as dizocilpine, paclitaxel, and some experimental oligonucleotides to a high end for HIV protease inhibitors (high doses) and some cardiovascular drugs (clinical studies of very large scope).
Source: Author's observations from involvement in numerous drug development programs.

at the same time for timely and successful product launch. Selected instances of such balancing, in which some risk is often inevitable, are discussed throughout the rest of the chapter; therein lies the crucial need for overall coordination of *each* drug's development program.

Although various models exist, today's drug development is generally facilitated by a coordination mechanism and forum, usually in the form of a cross-functional team that drives and manages a drug candidate. The principal objectives are to have and execute: (a) a drug development plan, (b) rigorous means to closely track its execution, and (c) mechanisms to effectively respond to events and findings that invariably arise in spite of the plan. Indeed, the development of a new drug encompasses a myriad activities and objectives that are extremely cross-linked among the various disciplines contributing to the effort. Clearly, the bulk process development team needs to be well represented in the cross-functional forum throughout the drug development cycle.

Success in development coordination means that, no matter which coordination model is used, there must be prompt and effective resolution of most issues and difficulties, say $> 90\%$, *at the team level*, with the rest going up to a broader and more senior team of the R&D organization (i.e., the heads of the disciplines, functions, and those above). Indeed, the direction and operation of such teams have become a distinct function (it will be referred herein as *drug coordination*), with its own set of skills and not unlike the distinct set of skills in new drug submissions and approval—the *regulatory affairs* function.

The relationships of the three basic tasks with the overall drug development program are depicted in Fig. 5 in rather simple terms, whereas the specifics of each relationship will be discussed under the heading of each task. The arrows indicate the flow of materials from the preparative task and the flow of information and know-how from each task to the others and to the drug development at large.

It is also useful to depict the bulk drug process development cycle on a Cartesian coordinate plane (Fig. 6). The

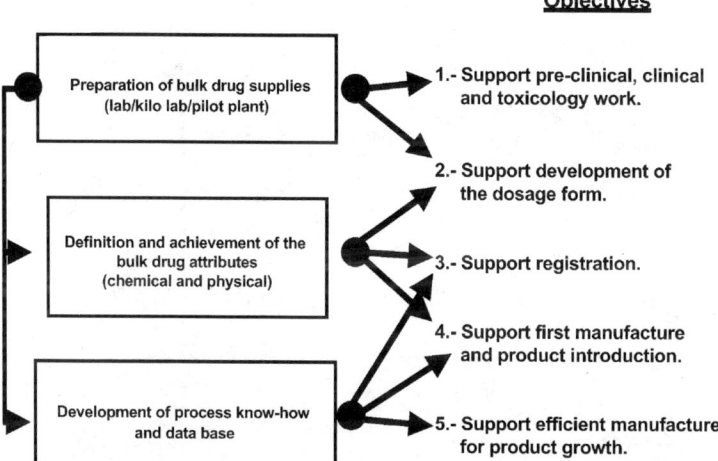

Figure 5 The three basic tasks of bulk drug process development. These tasks exist concurrently throughout most of the development cycle, albeit their burdens vary through the cycle. Nevertheless, managing well all three tasks as inseparable parts of a single overall endeavor is the principal managerial challenge in bulk drug process development.

abscissa axis represents progress since the onset of development of a compound and, although progress along well-defined milestones is used, one might also look at the abscissa as measuring the applied technical effort or, less precisely, the extent to which the bulk drug process has been reduced to practice (e.g., kilos of bulk drug made, batches made, or versions of the process piloted). Inevitably, the abscissa scale shown herein is arbitrary, albeit deliberately selected; the experienced reader will probably readily think of an example with a more apt progress scale. Thus, the need to deal with the latter in terms of more distinct *stages*, which Fig. 6 attempts to depict.

Were elapsed time to be used, the distance between *Phase II/III* start and the *Dossier* filing milestones would be quite variable from drug to drug, as that interval depends on the scope of the clinical program and on the therapeutic target. Whereas osteoporosis, diabetes, or depression require

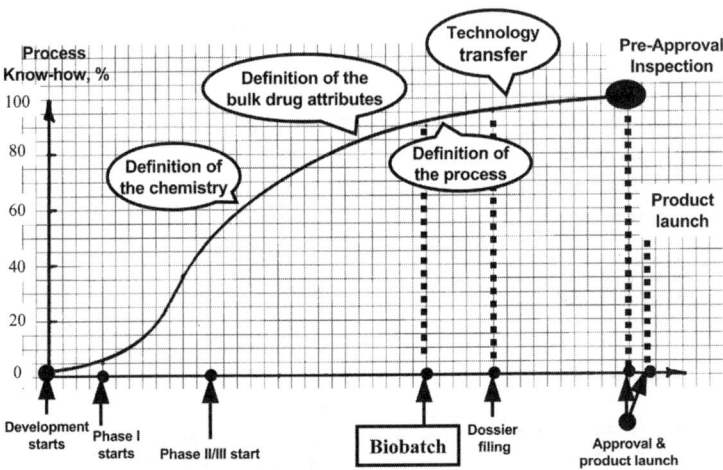

Figure 6 The process know-how vs. applied effort plane, including the major milestones of bulk drug process development. As defined herein, 100% know-how describes the body of knowledge needed for registration and reliable first manufacture for product launch, whereas additional know-how accumulates with manufacturing experience and follow-up work that might be done for process improvements or a second generation process.

considerable time to reach their efficacy endpoints, those for bacterial infection or pain relief, for example, arrive much sooner. For this, and for other reasons related to the intended scope of the drug development (e.g., claims structure, schedule of filings, multiple routes of administration, etc.), the elapsed time scale is unsuitable for the *process* know-how purposes of Fig. 6. Instead, applied effort or extent of reduction to practice of the process relate directly, if not strictly in direct proportion, to the acquisition of the process know-how.

Although the *biobatch* and *preapproval inspection* prerequisites are specific to USFDA approvals, analogous expectations are arising in other drug agencies in the major markets (more on this in Chapter 3). The *biobatch* is a distinct marker in dosage form development in that it serves as the bioavailability/bioequivalence bridge to pivotal clinical studies, as well as the bioavailability/bioequivalence reference for all subsequent dosage form output. As such, the biobatch reflects the process that goes into the dossier, *uses representative bulk drug* and excipients, and its size is no less than 10% of the intended manufacturing scale. *Preapproval inspection* is a methodology employed by the USFDA to ascertain, at its discretion, that the intended manufacture of dosage form and bulk drug correspond to the processes used in the pivotal clinical studies and described in the NDA or other new drug submissions.

The ordinate axis, on the other hand, is straightforward, as it measures the fractional bulk process know-how relative to that required for regulatory approvals and for sound first manufacture. Note, therefore, that it is not being suggested that at 100% on the ordinate axis there is nothing else to be learned about the process; instead, *the 100% ordinate value merely describes the knowledge required to fulfill the said process development objectives*. Indeed, further gains in process know-how are always realized with manufacturing experience, and mature processes often differ appreciably from their first manufacture versions, by virtue of gradual improvement or from significant step changes (*second-generation processes*), although most often the seeds for such later developments are planted in the original development body of knowledge. Thus, the curve in Fig. 6 describes the accumulation of know-how during four distinct phases of the process development effort:

a. The *preparative stage*, during which the effort is focused on making available kilogram amounts of the bulk drug to the preclinical, toxicology, and Phase I work, usually not based on the eventual synthesis route, let alone the eventual process.

Whereas the synthesis *route* (or *scheme*) describes the intermediate chemical structures sought to arrive at the final compound (starting materials, synthesis approach, and probable chemical reactions to use), the *process* describes how the route is implemented at a much higher level of detail. (solvents, catalysts, purifications, isolations vs. straight-through, etc.).

b. The *development stage*, in which the preparative work is scaled up and the synthesis effort goes into high gear, aimed at the manufacturing route *and* process. It is in this stage that the chemical engineering effort is applied in earnest, first to support the scaled-up preparative work and then to address the scale-up issues of the manufacturing route.

Ideally, the chemical engineering contribution starts early, so as to appropriately influence the seminal choices being made by the process chemists as to route. This influence is reasonably apparent with respect to issues of thermochemical safety and probable environmental impact; yet, there is across-the-board synergy that a chemistry/engineering dialogue can exploit. The latter is particularly true in those instances when the chemists perceive a desirable approach as not being feasible on grounds of scale-up difficulty or, more simply, because of lack of experience with some demanding processing conditions.

    c. The *consolidation stage*, in which the synthesis route is fully settled and the specific process for it is defined at the level of detail that permits process design for the manufacturing plant, definition of the bulk drug attributes and the assembly of the dossier. Also during this phase all the preliminaries for technology transfer are carried out and the stage set for first manufacture.

    d. The *technology transfer stage*, in which the process is run in its first manufacturing venue, its performance established, and the bulk drug needed for product launch is produced. Also during this phase, the manufacturing scheme receives approval within the approval of the dossier, often after plant inspection by the approving agencies.

From the above definitions, a discussion of the specifics of each stage is now possible, also based on the depiction of the bulk drug process development cycle on the know-how vs. applied effort plane introduced in Fig. 6. During these stage-specific discussions, the three bulk development tasks will serve as the basis and along the lines of Fig. 5.

C. The Stages of Bulk Drug Process Development

1. The Preparative Stage

Although preparative work takes place throughout the process development cycle, this first stage is most aptly described as the *preparative stage*. Its focus, although not exclusively, is the preparation of limited amounts of bulk drug for assorted preclinical purposes, then followed by first scale-up to support Phase I activities, which include testing the drug in healthy subjects (humans or the target animals if a veterinary drug).

Starting with bench scale equipment (up to 100 L in the so-called *kilo lab*) or pilot scale fermentors (up to 5000 L when titer is low), this early preparative work uses whatever synthetic method or fermentation conditions (the microorganism and the nutrients) are immediately available. In most cases of synthesis, the route may be a somewhat streamlined version of the discovery route or a temporary route that may or may not include parts of synthesis schemes being considered for eventual development. In most cases of biosynthesis, the microorganism is that from the discovery stage, but taken from whatever stage of microbial strain improvement is amenable to scale up from shake flasks or bench scale fermentors.

> Fermentation processes at this stage are generally of very low productivity (final concentrations of the target compound of $< 1g/L$), making access to relatively large fermentors most helpful, including, in cases of dire need, the use of manufacturing scale units (up to 75,000 L), the poor scaled-up performance of the early stage notwithstanding. The analogy for chemical synthesis is the arduous operation of lengthy procedures in the kilo lab, the low yields notwithstanding.

> Although the kilo lab will be described more fully later on, it may be said at this point that the kilo lab is a larger scale lab, traditionally used for running preparative procedures than for experimentation.

Preclinical and Phase I development work is crucial in that it determines the merit of further development or, hopefully, the adjustments that need to be made to move the compound forward. Thus, the importance of providing the required material on time to get those answers as soon as possible. This reflects on the need for capital investment in facilities such as kilo lab or pilot plant, and we will discuss elsewhere in this chapter the challenges of this stage of development when the preparative stage depends on *outsourcing* (the reliance on outside suppliers). Indeed, sufficient internal resources for the *preparative stage* is a clear competitive advantage, with the optimal setting providing the means—hardware and engineering skills—to swiftly overlap the kilo lab work with pilot plant work-up to, say, 1000 L vessels and the appropriate auxiliaries and operating environment (safety, industrial hygiene, and pollution abatement). Figure 7 depicts this preparative environment, whereas

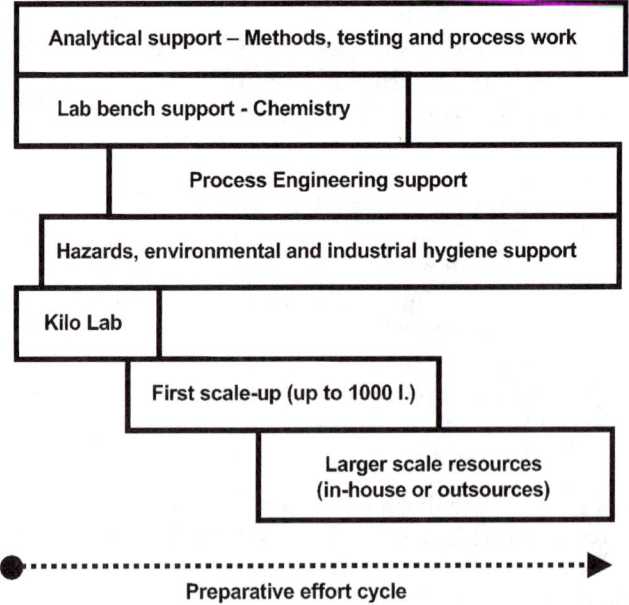

Figure 7 The resources for the preparative task. The need to engage larger-scale resources depends on the scope of the preparative task, which can vary widely (Table 1 and Fig. 8).

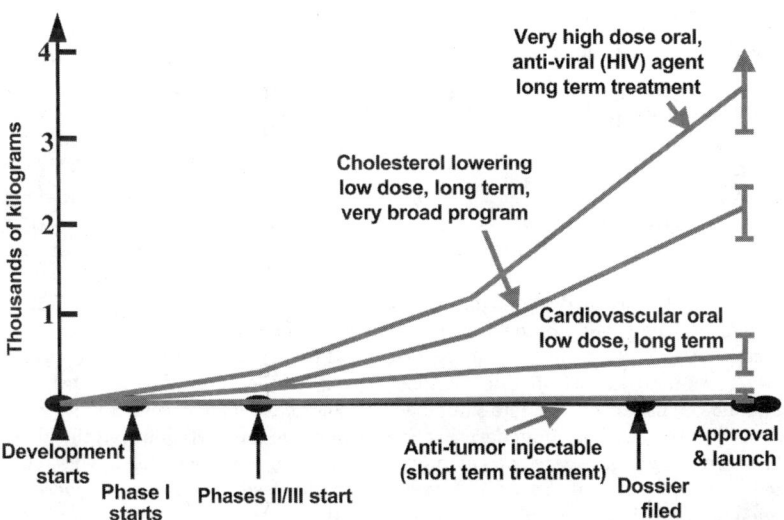

Figure 8 The scope of the preparative task. Some examples to illustrate the dependence of the preparative effort on drug potency, therapeutic target, and scope of the clinical effort.

Fig. 8 complements the range of preparative scopes presented in Table 1.

Also depending on the resources of the organization, synthesis bench work may take place in search of routes that can support a manufacturing process, as the routes used during the discovery phase are largely unsuitable on the basis of projected cost, length of the synthesis cycle, commercial unavailability of starting materials or simply because of their perceived inferiority relative to what the process chemists foresee as attractive alternatives. Clearly, the compelling wisdom of such early synthesis work needs to be balanced against the resources available and, most of all, against the empirical probability of less than 20% that a drug candidate at that stage will reach the market, as indicated by Table 2.

> Whereas medicinal chemists practice organic synthesis as an indispensable tool and are largely oriented upstream (towards the domain of biological and pharmaceutical attributes of the compounds they work with), process

Table 2 Best Practices Probabilities of a Drug Candidate
Reaching the Market

Drug candidates in the preclinical phase	5–10%
In Phase I	10–20%
In Phase II	30–60%
In Phase III	60–80%
Post NDA filing	> 95%

Notes: "Best practices" refers to drug development organizations with established good records of bringing drugs to market. In particular, best practices include a high hurdle for a drug candidate to enter development or Phase I.
Source: Author's assessment from assorted estimates, including those from the PhRMA Annual Report—online edition, 1997. While the figures from total compounds synthesized (or total number of biologically active compounds) have increased as the methods for generating actives improve their total output, the above figures *after entry into development* have remained largely unchanged. The above ranges probably reflect the adequacy of the tools used to assess the merit of developing an active compound and the rigor of the criteria for moving a compound forward.

chemists in the drug industry practice synthetic chemistry as their profession and are oriented downstream (towards the reduction to practice beyond their lab bench), thus the usual discontinuity in synthetic route at the discovery/development boundary.

Although sometimes much is made about smoothing and simplifying the discovery synthetic route (eliminating isolations and purification, shortening the processing cycle, and using less expensive materials), the most desirable contribution of the process chemist is the conception of a distinctly advantageous synthesis route that can then be developed and engineered into a sound manufacturing process. Such a route would bring the advantages of fewer steps from reasonably available starting materials, environmental benevolence (or, preferably, *green chemistry*), parallel moieties that can converge into shorter synthesis cycles, stereoselectivity, and similarly decisive gains.

As a summary, Fig. 9 focuses on the *preparative stage* and the rest of the preparative effort on the know-how vs. applied effort plane, whereas Fig. 10 depicts the materials flow from the bulk drug preparative effort at large.

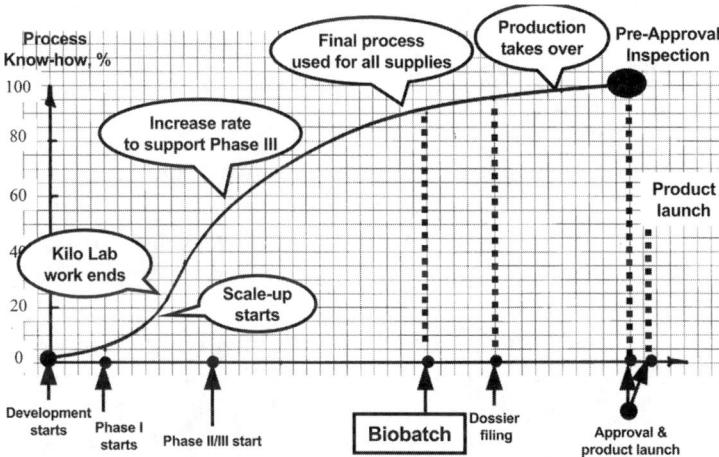

Figure 9 The preparative effort in the know-how vs. applied effort plane. The principal preparative milestones are shown.

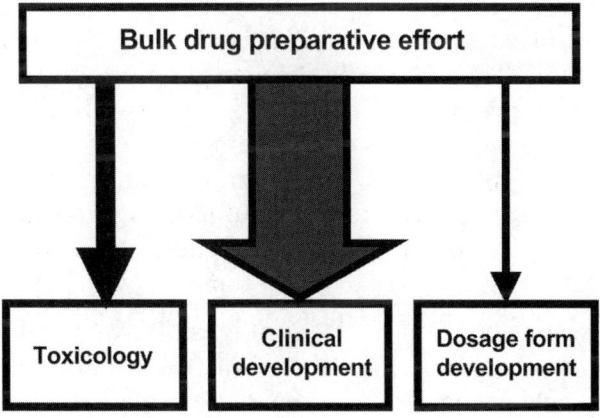

Figure 10 Materials flow from the bulk drug preparative effort. The width of the arrows *approximately* indicates the relative amounts of bulk drug going to the users in the overall drug development program. Examination of this figure and Fig. 9 provides an equally approximate description of the bulk drug usage as a function of the development cycle.

2. The Development Stage

As made clear by the slope of the curve in the know-how vs. applied effort plane (Fig. 6), the *development stage* comprises the most productive development effort:

(a) Synthesis work at the bench scale seeks the eventual manufacturing route in earnest, preferably on more than one approach, with all promising a substantial, if not overwhelming, advantage over the current preparative procedures.

In chemical synthesis, the route is basically driven by the structure of the target compound. Within that logic, however, the creativity of the process chemist is bounded only by the realities of starting materials availability. However, examples of bulk drugs made from commodity chemicals are now few and rapidly disappearing (thiabendazole and l-methyldopa, for example), as the more complex structures of today's medicinal chemistry preclude synthesis from basic raw materials. Instead, today's process chemist must be very alert to what the fine chemicals industry offers (or could be induced to offer) by way of suitable building blocks or intermediates and the corresponding manufacturing capabilities. Such alertness, combined with creative synthesis skills, is the key to truly advantageous routes. This theme is discussed amply and in depth in some of the previous references (2–5), as well as in Saunders's (11) compendium of selected major drugs. In the extreme, the total synthesis of structurally rich natural products, although rarely aimed at a manufacturing process, offers leads and inspiration to the process chemist, as well as blazes the trail with new reactions, some of which are eventually used in bulk drug syntheses (e.g., Ref. 12).

> In celebrating the opportunities for the creative process chemist, we should not neglect factors such as the increasing desire for environmentally benevolent chemistry (*green chemistry*) or the prevailing business model in the bulk drug industry, by which the range and scope of chemical processing has been narrowed in favor of contracting out (*outsourcing*). There is also, on management's part, the reluctance to practice hazardous chemistry

(nitration, sulfonation, phosgenation, etc.), with that spectrum of processing now all but ceded to contract manufacturers.

Some compounds of natural origin products have been manufactured by total synthesis when structurally simple (e.g., chloroamphenicol, fosfomycin) or when inevitable to bring a significant drug to market, as in the case of imipenem (13).

The selection of the chemical route, which is invariably made before it has been sufficiently reduced to practice, is the strategic decision, as it has the greatest potential to define the process and its overall performance—costs, reliability, environmental impact, etc. Accordingly, it is a decision that is best made with the benefit of sufficient engineering assessment, as sometimes the chemical appeal is not sufficient. Indeed, engineering assessments of capital and operating costs, environmental impact and issues of process design and scale-up bring sharply into focus the general direction as well as the specific development actions that the route requires to become the manufacturing process. On occasion, such assessments cause reappraisal of the route that, if timely, can redirect the project to considerable advantage— to a superior variation within the same basic route, or to a substantial change to a hybrid chemical scheme and, less frequently, to abandonment for another route.

Preferably, the synthesis route is settled not late during this stage, but it is not all that rare, in the higher caliber process efforts, for that "better route" to come through and displace the prevailing route just in time to switch the scaled-up preparative work.

It is at this stage of merging chemistry and engineering efforts that the process development effort generally settles onto the right track and the effort approaches critical mass. Process development organizations that lack the requisite engineering skills or that tap into relatively distant skills (say, from a technical resource in manufacturing) are at a marked disadvantage with respect to choosing the better process, since the said assessments are not done, are done less effectively, or done without the criticality of mass that the occasion demands. The distant engineering skills are also far less persuasive when their assessment of the proposed synthesis is not favorable.

All seasoned practitioners of bulk drug process development know from at least one experience the very high price paid when the wrong process gets too far down the development cycle, and retreat is either unacceptable or very costly to the overall development timetable. Thus, the compelling need to make the fundamental choices of route, and of process approach within the route, with the full set of skills and address the key questions:

1. What will the commercial plant look like? What will its operation be like?
2. What are the probable capitals costs? How long will it take to be ready to start up?
3. What are the scale-up issues? Can they be addressed on time?
4. What is the environmental impact? Is there a good fit with the likely plant sites?

Once the bulk process team gets past this juncture with an action plan, the rest of the development stage is mostly a matter of good execution by all the disciplines involved. Although the Analytical R&D function has not been mentioned up to now, its role is, of course, pervasive throughout; first in support of the early preparative work (a duty that remains with the function for the rest of the development cycle), then in decisive and indispensable participation of the development activity at the bench and in the pilot plant.

Biosynthesis processes, which are based on fermentation processing in which the microorganism does the synthesis, face the same set of development issues, but in a narrower field of options. Not only is the biosynthesis well defined and fixed by the microorganism, but alternate microorganisms with radically different pathways that could be more desirable are not that available. Chemical entities of natural origin are secondary metabolites of microorganisms or plant cells, and variations in the metabolic pathways that lead to a given secondary metabolite are relatively narrow compared to the many variations by which a compound can be made by chemical synthesis.

In this case, the development team (microbiology and biochemical engineering) aims at coaxing the organism or plant cell to be more effective. Strain mutation is a proven technique for improvement of the productivity of microbial biosynthesis. Plant cell processes, although very few in industrial practice, also seem amenable to increased productivity by manipulation of the cell lines and fermentation conditions. The microbiologist and the biochemical engineer are thus able to offer the potential for increased fermentation output by factors up to an order of magnitude or more—a potential not to be matched by increased yields from an organic synthesis. Indeed, some fermentation processes can go into manufacture at low titers with a high probability that increases will be obtained with continued development of the microbe or plant cell, as well as the fermentation conditions. Thus, variations on the biosynthesis—unlike variations on how to chemically synthesize a compound—are modest in range, but not in significance to fermentation productivity (e.g., use of phenylacetic acid as a precursor in the fermentation of penicillin G) or other important aspect of the process (e.g., switching to a different *Taxus* plant from which a precursor to paclitaxel, comprising the taxane ring with all of the desired stereochemistry, could be extracted and chemically converted to paclitaxel at an advantage over the prior extraction of paclitaxel).

It is in the processing downstream of the fermentor that development possibilities become numerous, as a wide range of unit operations for concentration, purification, and isolation exists, just as wide as the processing options for recovering the desired compounds from streams (i.e., materials) issuing from chemical synthesis. This is discussed much further elsewhere in this chapter.

(b) It is also in the *development* stage that the preparative work is scaled up in earnest with two purposes: (1) greater output of bulk drug and (2) the identification and resolution of the problems of scale attendant to the desired process. Although the latter goal requires that the desired route be at the scaled-up stage, considerable progress can be made if pieces of the desired route are scaled-up before the total route is brought to the pilot plant.

(c)   It is also during the development scale that the definition and achievement of the desired physicochemical attributes of the bulk drug is pursued in earnest, hopefully after the dosage form development team has narrowed down the ranges for those properties after the major decision—which particular salt or the free base or the acid will be the bulk drug form of the biologically active structure. Such a decision may come late in the cycle, for oral drugs in particular, as the search for the desired bioavailability and stability may be arduous (14).

(d)   Finally, it is during the *development stage*, preferably early, that the bulk development team starts its work with the appropriate downstream organization in anticipation of successful drug development, registration, and market launch. This set of activities takes place in a rather distinct track from the R&D track, often placing inordinate demands on the bulk process team, as their obligations to the drug development effort remain unaltered by the onset of their obligations to eventual technology transfer.

> There is a great deal of risk when bulk process resources are badly caught in the vise of the demands from their drug development partners and the increasing demands of technology transfer. Staffing of the bulk process team— the engineers in particular—needs to recognize that successful drug development brings with it technology transfer. Unfortunately, R&D management and the peers in the drug development program are often insensitive or oblivious to the situation and the cross-functional coordination team needs to be indoctrinated accordingly. It is very helpful to have the downstream functions related to manufacturing participate in the coordination team and thus ensure that those demands get known, if not fully appreciated.

In summary, Fig. 11 depicts the *development stage* in the now familiar know-how vs. applied effort plane. It is also

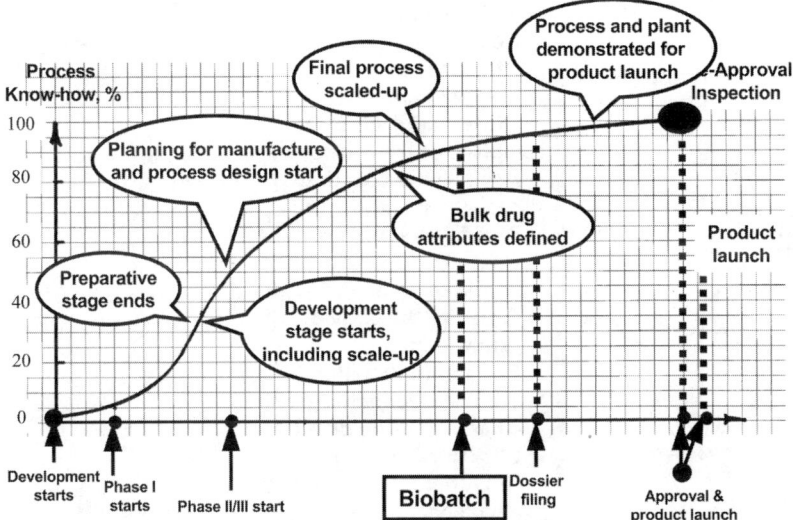

Figure 11 The development stage in the know-how vs. applied effort plane. The principal process development milestones are shown.

timely to present the full spectrum of the bulk drug develop-ment disciplines and all the activities that they carry out, including those shared with others in the corporation or with outsources, as shown in Fig. 12.

3. The Consolidation Stage

Although it is not infrequent for a significant bulk process "loose end" to remain tenaciously loose until late in the cycle, by and large the development cycle reaches a stage at which the more difficult development work has been done. To wit:

    a. The chemical synthesis route is fully defined, albeit sources and specifications of starting materials may still be under negotiation or definition.

    b. The actual process based on the synthesis route is sufficiently defined and sound pilot plant operating procedures exist or are clearly in the offing.

    c. Preparative support to the drug development program, although continuing and never leisurely,

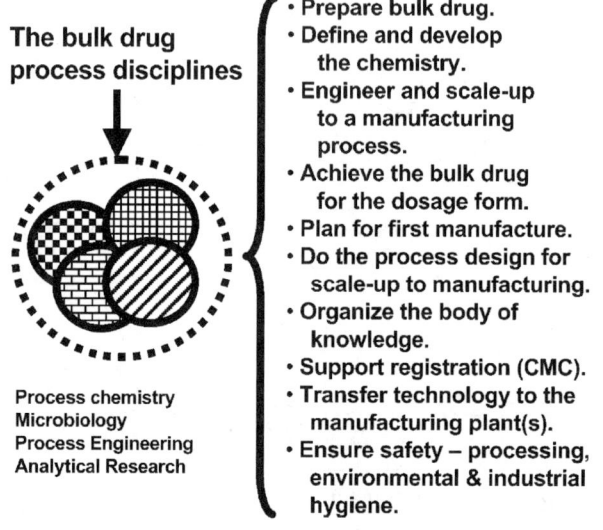

The bulk drug process disciplines

- Prepare bulk drug.
- Define and develop the chemistry.
- Engineer and scale-up to a manufacturing process.
- Achieve the bulk drug for the dosage form.
- Plan for first manufacture.
- Do the process design for scale-up to manufacturing.
- Organize the body of knowledge.
- Support registration (CMC).
- Transfer technology to the manufacturing plant(s).
- Ensure safety – processing, environmental & industrial hygiene.

Process chemistry
Microbiology
Process Engineering
Analytical Research

Figure 12 Disciplines and activities in bulk drug process development. CMC stands for the chemistry, manufacturing and control component of the dossier.

is no longer threatened by uncertainties about how to prepare the bulk drug.

d. Thermochemical safety data are firm and only updating for process changes remain to be done. All issues are being dealt adequately in the process design of the manufacturing plant.

e. The environmental impact of the process at the site of manufacture and at large is understood and acceptable, meeting company policy objectives. Obtaining all the requisite permits is likely.

f. Industrial hygiene issues specific to the process are understood and being addressed adequately in the process design of the manufacturing plant.

g. The process design, and possibly plant construction, are proceeding. Uncertainties seen within the grasp of the combined development/process design effort and work can be focused accordingly.

    h.   Analytical methods for in-process and bulk drug control have been largely defined and remain to be confirmed and validated. Absolute purity, impurity profile, and crystal form are settled matters.

    i.   The scope and approaches to the dossier are largely in hand, if not in text.

There is, of course, no suggestion of the work being completed. Far from it, the *consolidation stage* is intense in a different way that the *development stage* was. A great deal of the work ahead is filling blanks (few if the prior work has been done well), refining pilot plant procedures and catching up on the documentation that will support the dossier. Also, the final work on the definition and achievement of the bulk attributes needs to be done to support the final work on the dosage form side and the biobatch and stability studies that will follow.

There is also the largely increased workload in preparation for technology transfer, usually requiring frequent travel, a great deal of interaction, and the pursuit of much detail. Snags in process design and plant construction do come up, environmental permits may require scrambling for some data, etc.

However, the slope of the know-how curve is decreasing rapidly, as the bulk process is being implemented more than it is being developed, the loose ends notwithstanding. In summary, Fig. 13 depicts the *consolidation phase* in the know-how vs. applied effort plane.

4. The Technology Transfer Stage

Most of the discussion on the nature and scope of the technology transfer activity is presented in Chapter 3. Nevertheless, the following seems pertinent at this point, as it relates to the technology transfer burden that the bulk process development team carries in addition to its duties on the drug development program:

    a.   A finite effort, even if the midst of a very difficult *development stage*, must be allocated to look ahead

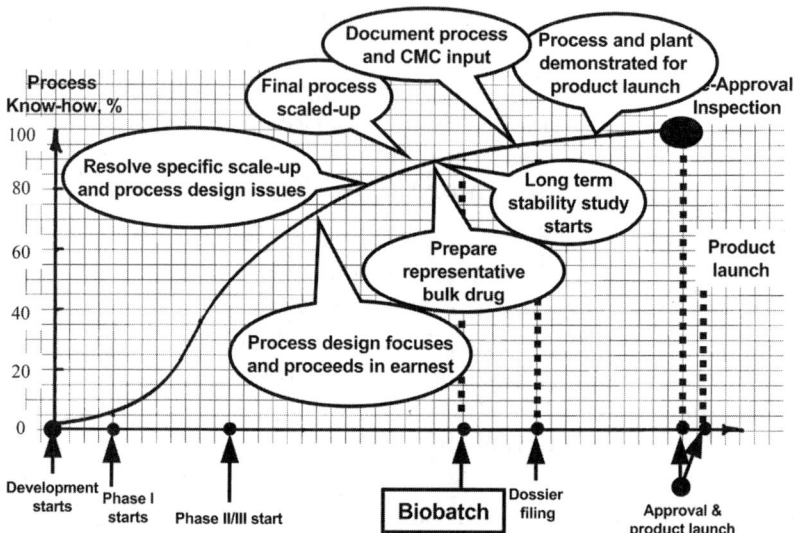

Figure 13 The consolidation stage in the know-how vs. applied effort plane.

to the specifics of manufacturing the bulk drug. This has been indicated in Figs. 11 and 13 .

b. The bulk process team needs to keep the rest of the R&D organization, their peers in the coordination team in particular, aware of this downstream task.

c. The technology transfer team needs to be well rounded—chemists or microbiologists, engineers and analysts—and at the site of technology transfer. Staffing and briefs to do the job should be generous to decisively start up the process for product launch. No rescue missions allowed!

d. Successful technology transfer—from early planning for manufacture, process, and plant design, process start-up preliminaries and the actual demonstration that the process works in the commercial plant— rests squarely on the process body of knowledge being as complete as needed by the task and organized to effectively impart knowledge to the downstream organization.

    e. Regardless of what organizational arrangement might exist, the bulk process development team needs to assume, hopefully in a collaborative understanding, a leadership role as the bringers of the know-how.

    f. With the necessary adjustments, all of the above apply when transferring the process technology to contract manufacturers or licensees. More on this under VI—Outsourcing in bulk drug processing.

III. FROM THE BENCH TO THE PILOT PLANT AND BEYOND

A. Process Conception and Bench Scale Development

Except for fermentation or recovery from natural sources, all other chemical entities are obtained by chemical synthesis from organic chemicals and the process conception starts with that of the synthesis route—the scheme by which selected starting structures are converted to the target drug candidate. Factors considered by the synthetic chemistry team are:

(a) Starting materials that are available (or could be available) that promise an attractive route, and a *wish list* for such a route could be as follows:

1. The route is direct, with few steps needed to reach the target compound.
2. It is also convergent (two moieties can be assembled in parallel, then joined near or at the target compound), thus offering shorter synthesis cycles and higher yields.
3. If chirality is sought, it appears attainable through enantioselective methods.
4. Once chirality is obtained, the route preserved it.
5. Minimal need for blocking/deblocking.
6. Absence of highly hazardous materials, reactions, or intermediates.
7. Environmentally benevolent (i.e., green chemistry).

8. Probable cost is appropriate to the product.
9. Fits nicely with existing plant running a related process.

The relative priorities of these factors vary widely, as they are seldom all present; neither are they fully independent from each other. For example, directness of synthesis may come at the price of a very expensive reactant, or would require that a very hazardous intermediate be made and perhaps isolated. Or perhaps the greenest route seems least feasible. Additionally, the selection may be constrained by compelling demands of the drug development program: for example, the most attractive route would take longer to be ready for preparative work and development; it has to defer to the lesser route that can prepare bulk drug now—not an uncommon juncture and decision, although it can be subsequently reversed.

Indeed, there is no established system to deliver the best or even a very good choice of synthesis route, and creativity and synthesis acumen still dominate, although obviously aided by the above and other simpler criteria, such as that of "atom economy" (how many atoms of the reactants end in the final compound?) (15). Occasionally, the choice is facilitated by a compelling case of an ideal starting material availability (e.g., a chiral intermediate that would bring all or a good deal of the target chirality with it), a selling approach that fine chemical producers exploit. Then at some point soon, the leading choice of route needs to be challenged by the various engineering assessments described in II. C.2.a.

Bench development of the route (or routes) of choice is pursued aggressively, ideally by both synthesis chemists and chemical engineers, with the former elucidating reaction pathways and byproducts, seeking superior reaction conditions (solvents, catalysts, auxiliary chemicals, temperature, pressure, concentrations, reactant ratios, and approximate kinetics) as well as probing work-up and isolation methods. The engineers work, in collaboration with the chemists, on aspects of the chemistry better suited to their skills (e.g., kinetics and thermochemistry, multiphasic reaction systems with mass transport effects that distort the chemistry, or very fast reaction with selectivity issues that are sensitive to mixing, or reactions requiring concurrent separation or continuous reactors with tight control of residence time or extraordinary heat removal provisions, etc.).

Such bench development by both disciplines is what transforms a synthesis route into a process candidate for scale-up and eventual manufacture. If done concurrently—as it should be—it allows for the results to flow across the disciplinary boundary and shortening the path to a sound process derived from a sound choice of route.

(b) Fermentation or natural product extraction processes, on the other hand, are not burdened by a broad range of route possibilities, as discussed under II.C.2.a. Bench development by microbiologists and engineers, however, is indeed rich with possibilities. To wit:

For microbial or plant cell fermentations:

1. elucidation of the pathway to the secondary metabolite;
2. nutrient, precursors, and optimization of fermentation cycle conditions (from the above results);
3. strain and cell line improvements with respect to productivity and robustness in fermentation;
4. data gathering to support scale-up to stirred tanks at all pilot plant scales;
5. definition of the downstream process candidate for recovery, concentration, purification, and isolation of the target product from the fermentation.

For extraction of compounds from natural sources (plant or animal material):

1. evaluation of differing sources of the compound bearing materials;
2. pretreatment conditions for successful extraction;
3. extraction or leaching conditions, solvent, or extracting stream (i.e., material) selection, and separation of spent plant material;

    4. definition of the process candidate for concentration, purification, and isolation;

    5. data gathering to support scale-up.

Most likely, both technologies eventually have to deal with relatively large volumes of cell mass or plant material waste, and bench work to address those issues is also needed.

B. Process Scale-Up

1. What Is Scale-Up?

At its simplest, scale-up is the set of processing issues that arise when the same operations take longer to execute in larger scale equipment than at the bench scale. Although such issues do arise, they can be anticipated and in most cases avoided or largely mitigated through changes to the design and operation at the larger scale.

Much more often and less apparent, however, are the processing issues created by operating at larger scale—*with greater dimensions and different geometries*—and thus affecting flow regimes, phase separation rates, interfacial surface areas, mass and heat transfer rates, flow patterns, heterogeneity in process streams (i.e., materials) and many other dimensionally sensitive variables and parameters. These effects are not related to a different time scale of processing events, but arise instead from strictly physical effects that distort the process results from those at the small-scale baseline, including chemical outcomes. Relevant examples are:

    a. Reactants to a system of fast reactions cannot be mixed fast enough and fractions of the reaction mass proceed for finite times at concentrations very different from the intended average concentration (some fractions unduly rich in the reactant being added, while others are unduly low), resulting in a product distribution different from that predicted by the kinetics or obtained at the smaller scale.

    b. Mixing in larger stirred tanks, if not adjusted properly, can result in significant differences in the composition of matter of multiphase process masses across the tank volume relative to the more uniform results in smaller tanks.

c. Rotating devices of larger diameter, such as agitators and pump impellers, as well as internal moving parts in a solids mill, will exhibit higher tip linear velocities and thus generate greater shear stresses in fluids or contribute greater energy to impacts relative to the analogous operation at the smaller scale.

d. Crystallization processes at larger scale can suffer from unwanted nucleation as the result of heterogeneities in solvent phase composition during semibatch addition or in local temperatures upon cooling, as well as more prone to crystal attrition and contact nucleation from the higher tip speed of the agitators and greater energy impacts among particles.

e. Transfer rates of sparingly soluble gases into liquids in stirred tanks generally suffer with increasing scale of the tank unless provisions are made to mitigate the differences, as the gas bubble size distribution (and with it the interfacial surface area) generated by the agitator impeller is different. Hydrogenation rates observed in laboratory pressure vessels, for example, most often do not scale up to pilot scale stirred tanks because of the extraordinary gas absorption obtained in the liquid vortex at the lab scale; the larger pilot scale tank, being equipped with baffles, does not generate a vortex and that contribution to gas absorption is not present.

f. Large process vessels lose heat less rapidly than smaller vessels at the same internal and ambient conditions and, when deliberately cooled, will cool less effectively, absent a mitigating cooling provision.

g. Larger continuous contacting vessels for devices for gas/liquid, vapor/liquid, solid/liquid, and liquid/liquid systems will perform less well because of maldistribution and bypassing of the phases worsens as the cross-sectional area of the contacting vessel increases. Such scale-up requires that provisions be made with internal parts to alleviate maldistribution.

h.   Flow vessels will exhibit different flow patterns and residence time distributions than smaller vessels, which need to be taken into account so as to design the larger vessel accordingly.

Indeed, carrying out a processing operation at a sufficiently larger scale often shifts the rate controlling step of the process event *from one domain to another*. As an example, in reactions in gas/liquid systems the small scale usually permits the reactant in the gas phase to be abundantly available to the liquid phase (the rate of chemical kinetics is observed, *as the gas/liquid mass transfer is not limiting*); whereas upon scaling up *the gas/liquid transfer may become limiting* and the reaction, now starved for the reactant being supplied by the gas phase, does not follow its expected kinetics. The result of such shifts may go beyond the different rates of reaction, as selectivity (and relative rates of impurities formation) may change upon lack of a reactant. Generally, chemical reaction systems that have very fast rates or that take place in multiphase systems are sensitive to the operating scale due to the intrusion of mass transfer effects upon the performance of the chemical kinetics.

The above is a partial list of frequent scale-up issues that arise in bulk drug processing with consequences of lower chemical yields, or worse yet, loss of control over the impurity profile, as well as slower processing, excessive damage to microbial cells and crystalline solids, undesirable particle size distributions and any from a wide range of assorted shortfalls in process performance.

Understanding, *predicting*, and dealing with these issues requires more than a modicum of chemical engineering skills, such as fluid mechanics, mass and heat transport, the use of dimensional analysis tools, and mathematical methods for the simulation of events in a new context. Absent those skills, scaling-up will result in surprises, cause much less effective trouble-shooting, and engender an unwarranted fear of scaling-up. Indeed, such apprehensions are now codified in arbitrary batch size ratios beyond which regulatory constraints to process change apply.

Often enough scale-up is done much too tentatively, inserting intermediate scales that are not needed. Direct scale-up from the lab to the plant is quite feasible in a number of cases (e.g., fast liquid phase reactions with known kinetics and thermochemistry). All that is required is that the issues be understood and the proper parameters reproduced or improved at the large scale, using adjusted process conditions, as it is the set of the defining parameters what needs to be reproduced, not necessarily each process condition.

Failure to understand scale-up issues equates a change in scale with a change in the process. While it is appropriate for a change in operating scale to come under the scrutiny of a well-managed change control system, there should be no assumption that it is "the process" that is being changed; a distinction that is not about semantics, but about the approach to scale-up by the practitioner. This pertains in particular to operation of a pilot plant, in which scaling up and changing the process are a daily overlap that, if not practiced with a sufficient understanding of what is happening, will often befuddle the practitioner.

Yet, scale-up is inevitable, even in the relatively low throughput environment of bulk drugs. Skillful use of the pilot plant environment, by which the preparative task and the process development scale-up coincide in time and place, is essential to a vigorous bulk development program lest the activity oscillate between the extremes of unskilled scale-up and fear of scale-up. Indeed, lack of sufficient scale-up skills is a major disadvantage in bulk drug process development.

2. Tools for Scaling Up

In addition to the engineering skills and the access to the full range of supporting laboratory capabilities (bench development, in-process, analytical, physical chemistry, microbiology), scaling-up requires a variety of measurement apparatus (e.g., compressibility cell to measure flows through beds of solids at different compression), as well as the frequent assembly of dedicated apparatus or pilot units (e.g., units to measure fouling rates of surfaces over short-term test, small-scale centrifuges to more reliable measure centrifugation rates, leaf test units for vacuum filtration tests). It so happens that often enough studies and measurements cannot be made in processing equipment nearly as well as they can be made in a smaller-scale apparatus dedicated for the purpose at hand. The enterprising scale-up team will, in due course, assemble and accumulate such test apparatus as the needs arise.

In addition, some scale-up works need apparatus that are operated for preparative purposes as well, along the lines of the kilo lab, but in a flexible environment not focused exclusively on batch processing as the kilo lab is. Examples of such apparatus are fluid bed crystallizers, hydroclones for the evaluation of that method of solid/liquid separation, lyophilization cabinets with special vial sampling capabilities, intermediate scale membrane processing assemblies, etc. An area well suited for such testing purposes is not only highly desirable, but often facilitates preparative work by processing methods not within the scope of the kilo lab. Such an area should be reasonably open for the manipulation of portable equipment, with ample walk-in hoods and tall California racks, well distributed utilities, portable measurement panels for recorders, flowmeters and the like.

C. The Pilot Plant and Its Objectives

The objectives of the pilot plant environment in bulk drug process development are simple:

1. provide ready access to scaled-up processing for the bulk drug preparative effort;
2. give the preparative effort a responsive environment to deal with the vicissitudes of the drug development programs being supported;
3. permit the rapid and convenient evaluation of processes, methods and equipment, as well as the mastery of their scale-up;
4. obtain process data to support process design and the development of procedures for the eventual manufacture of the drug;

5. demonstrate process performance at a scale that minimizes risk and the need to trigger scale-up constraints upon first manufacture.

Of the above, only the last seems to require elaboration at this point, as process design is amply discussed in Chapter 3. Thus, given a sufficiently large scale of processing in the pilot environment where the development effort takes place, the transfer to manufacture will be far less likely to entail scale-up risk and, more importantly, far less likely to create a regulatory scale-up issue upon first manufacture. Bulk drug pilot plants of recent construction at R&D drug companies and at the major contract manufacturers provide batch processing vessels up to 10,000 L for chemical work and up to 20,000 L for fermentation work.

Table 3 outlines the equipment capabilities for a broadly capable bulk drug pilot plant, such as can be found in the major drug R&D companies, albeit not necessarily all in the same location. Such plants have generally resulted by accretion and as needed to support vigorous drug development programs; say, more than 10–15 compounds among preclinical and all phases of clinical work. Indeed, pilot plant capabilities of similarly broad scope can also be found in the premier contract manufacturing companies, as they are active participants in the preparative work for the whole spectrum of drug R&D companies.

Although in the 2000s the outsourcing field is populated with a great many small firms claiming to have cGMP pilot plant capabilities, Table 3 and the desired capabilities listed just below make it clear that the drug development business is one in which size does matter. Obviously, there is a sliding scale of capabilities vs. scope of physical plant, and Table 3 describes its full range so as to *decisively* accomplish all of the above aspects of pilot plant work for bulk drugs. Lesser options entail lesser preparative power, narrower range of scale-up and processing technologies, and processes tailored to fit existing capabilities. Alternatively, a combination of outsourcing and in-house resources can be used, but most often with far less agility of preparation and development.

Physical plant by accretion means that, short of large
lumps of capital investment, the processing capabilities
will span designs and practices evolved over decades.
However, issues of industrial hygiene and regulatory
expectations have gradually done away with open processing areas and rows of vessels in favor of processing modules in various degrees of segregation and connectivities.
Figure 14 describes one such prototype of modular design
for bulk drug chemical processing.

This table attempts to list most, if not all, of the equipment
for a *comprehensive* bulk drug pilot plant. Accordingly, it is a
wish list requiring a great deal of capital investment for fulfillment. Nevertheless, such pilot plant "complexes" do exist, arising mostly by accretion over decades, but also from projects of
large scope used to augment and modernize earlier physical
plant obtained by accretion. Pilot plant investment is generally
viewed by R&D management as strategic, reflecting their long-
term assessment of the vigor of the new drug pipeline, as well
as their unwillingness to accept preparative bottlenecks.

Some capabilities are needed very infrequently and are
best secured through vendors, other companies, or universities
that might possess them. For example: molecular distillation,
high vacuum fractionation, gas/solid elutriation at scale, fluidized bed crystallization, gas/solid catalytic reactors, vacuum
belt filters, gas/solid ball mill reactors (a la Kolbe), very high
pressure stirred autoclaves and similar equipment. However,
some projects reach the point at which the case for the in-house
capability becomes compelling, particularly if the technology
at hand seems attractive at large. Examples of such technologies are: fluidized bed crystallization for the resolution of enan-
tiomers, gas/solid catalytic reactors for selective oxidation or
dehydrogenation of heterocycles, permeable wall tubular reactors and membrane-aided liquid/liquid extraction.

Similarly, hightly specialized capabilities and new technologies are best provided at an intermediate bench/pilot scale
for ready evaluation of whatever advantage might drive the
larger scale proposition. This is a large part of the rationale
for the "bench/pilot" processing area, ninth in the list below.

For the smaller organizations, the appropriate pilot plant capability presents a formidable challenge. Preparative power and the ability to handle varied and often unpredictable processing tasks hinge on owning or having ready access to a sufficient breadth of processing equipment, preferably in a context amenable to experimentation at scale. Absent these capabilities, the processes so developed are, inevitably, highly constrained in their scope and technical ambition; compromises are inevitable and timeliness and assurance of bulk drug preparation trump any other consideration. As to outsourcing as an adequate complement, its limitations are often severe, a principal topic discussed under VII.

Table 3 Physical Plant of the Comprehensive Pilot Plant for Bulk Drug Processing

---

Infrastructure
 Warehouse space—partitioned and protected to meet cGMP and safety requirements
 Materials handling suitable to the scale and scope of the processing tasks
 Utilities systems or suitable distribution from site systems
  Steam (up to 11 atm)
  Cooling water (variable temperature depending on the source and season)
  Chilled water (down to 5°C)
  Refrigerated coolant (down to −25°C)
  Compressed air (up to 5 atm)
  Nitrogen (up to 2 atm)
  Portable hot fluid recirculating system (up to 250°C)
  Portable cryogenic recirculating system (down to −100°C)
  Electrical power (AC of normal voltage and of voltage required by industrial motors; e.g., 110–440 V in the United States)
  Fire protection (no less than that required by the applicable codes)
  HVAC (rather variable according to space being ventilated)
  Tank farm (solvents, acids, bases)—all above ground, dikes as required
Pollution abatement systems (may vary widely in scope according to site circumstances)
 Acid/base neutralization capabilities (at source)
 Scrubbers (water and aqueous base) for the processing areas
 Carbon adsorption systems for specific emission points
 Trim condensers as required in processing vents
 Thermal oxidizer and stack for process vent emissions
 Tanks (segregated by waste category, dikes as required)

(Continued)

Table 3 Physical Plant of the Comprehensive Pilot Plant for Bulk
Drug Processing (*Continued*)

Ducting and fans for tie-ins to the various abatement systems
Hazardous waste storage and unloading to haulers
Chemical processing areas: Nine distinct processing areas should be
  considered during the design or longer term planning of a
  comprehensive pilot plant for bulk drugs:

1. *General chemical synthesis processing.* Processing duties that do not
   fall squarely into any of the areas below
2. *Finishing area for bulk drugs (includes solids processing).* Light
   chemical processing, if at all. Key duties of this area are to crys-
   tallize the final compound as the bulk drug (with the desired
   chemical and physicochemical attributes). Includes filtration, drying,
   milling, classification, compacting, blending and packaging. Environ
   ment is distinctly cleaner than most other areas
3. *Aseptic finishing area for bulk drugs (includes solids processing).* No
   chemical processing other than salt formation. Includes all of the
   above finishing area provisions, but largely in an aseptic processing
   environment for the preparation of sterile bulk drugs. Requires
   sterilization equipment, special ventilation systems and much greater
   partitioning of the space
4. *Hazardous processing for toxics, hydrogen, nitration, sulfonation, etc.*
   Segregated operating space with extraordinary fire, explosion venting
   and ventilation provisions. Contains the key equipment for reaction
   and limited work-up
5. *Highly potent compounds processing.* General processing equipment in
   a segregated area and equipped for a high degree of containment of
   materials being handled, due to industrial hygiene and environmental safety reasons
6. *Housekeeping (neutralization and other disposal activities with waste
   streams).* Complements the above areas, sometimes being within or
   adjacent (e.g., Fig. 14)
7. *Fermentation processing.* Very distinct in space, equipment and
   auxiliary facilities (16). Microbiology lab, seed development lab, and
   fermentor train. Air compression, air and liquids sterilization, tank
   and piping sterilization. Stirred tank fermentors, feed tanks and
   harvest tanks
8. *Downstream processing of fermentation streams.* Also a very distinct
   processing area: little chemistry but a great deal of work-up,
   purification and isolation with a different mix of unit operations (17)
9. *Intermediate bench/pilot scale lab for engineering studies* (not the kilo
   lab, although it can be readily pressed into preparative duty as
   appropriate). Multilevel open bay space, walk-in hoods, tall racks,
   utilities stations for rented portable equipment, very little fixed
   equipment

(*Continued*)

Table 3 Physical Plant of the Comprehensive Pilot Plant for Bulk Drug Processing (*Continued*)

Clearly, each of these areas has different requirements, but it is not in the scope of this chapter to attempt a discussion beyond the above outlines (23).

Processing equipment

General purpose stirred vessels in the 100–5000 L range. Glass lined/316L stainless steel/specialty alloy in an approximate ratio of 1/0.4/0.1 in frequency. Vessels above 100 L should have split jackets. All vessels intended for a processing function (as opposed to waste neutralization, solution make-ups, etc.) should have variable speed drives for their centerline agitators and be baffled accordingly. Vessel layout and connectivity can vary widely, but organization into multipurpose once-through gravity modules seems to be the most useful for piloting purposes (i.e., from top to bottom levels: set-up, reaction, work-up, crystallization, and solids/liquid processing and housekeeping). Please refer to Fig. 14

Within the category of general purpose vessels, a variety exists that the design can put to good use. For example, vessels intended for crystallization will often have agitator impellers better suited for that purpose, or vessels intended for work-up of reaction outputs will often be fitted with auxiliary devices for liquid/liquid extraction or for evaporative concentration.

Fixed auxiliary equipment for general purpose stirred vessels (sized accordingly): condensers, decanters, receivers, weighing tanks, solids charging devices, sampling devices, pumps and piping, connectivity to the tank farm and to pollution abatement equipment, connectivity among each other, vacuum sources, vent trim condensers, overhead catch tanks, in-line filters, flow splitters, etc.

Portable auxiliary equipment for general purpose processing: pumps of various kinds (centrifugal, positive displacement, vacuum), drum handling devices with pumping provisions for charging to vessels, scales of various ranges, stirred tanks (slant agitator) in the 100–1000 L range, recirculating sampling or sensor loops, small filter press or pressure plate filters, blow charge tanks, line mixers, etc.

Specific processing equipment and their auxiliary equipment (as above):

a. *High-pressure reaction stirred vessels* (glass lined up to 5 atm, 316L stainless or specialty alloy to 15 atm).

b. *Liquid/liquid extraction and phase separation devices*: centrifugally aided, including those capable to deal with suspended solids (i.e., for fermentation broths processing); mixer-settlers, membrane coalesc-

(Continued)

Table 3 Physical Plant of the Comprehensive Pilot Plant for Bulk Drug Processing (*Continued*)

ing filters, rotating internal columns, etc.

c. *Gas/liquid contacting devices* (other than pollution abatement devices): Overhead venturi contactors, high turbulence contactors, packed or tray columns, wetted wall columns, etc.

d. *Evaporation and distillation equipment*: falling film, long tube, and wiped film evaporators with their condensing, receiving and vacuum sources; fractional distillation columns with their accessories, vapor/liquid disengagement inserts in selected vessels, etc.

e. *Adsorptive processing equipment*: columns and accessories for ion exchange, chromatography and other solid/liquid adsorption, high-performance liquid chromatography systems (columns, tankage, influent delivery devices, sensors and controls), molecular sieve solvent dryers, etc.

f. *Solid/liquid separation devices*: centrifuges (center-slung dig-out (up to 24″) or bottom drop (up to 48″), horizontal axis and side discharge), pressure and vacuum filters (stacked disk, plate and frame, agitated filter-dryers), cross-flow filtration, polishing filters, etc.

g. *Solids drying equipment*: tray dryers (vacuum and air), fluid bed dryers, vacuum tumble dryers (with assorted internals), stirred filter/dryers, countercurrent solids/gas dryers, spray dryers, lyophilization systems, etc.

h. *Solids processing equipment*: fluid bed and rotating shell processors, assorted grinders and mills, compactors and extruders, classifiers and blenders.

i. *Solids/liquid processing equipment*: homogenizers, colloid and ball mills, fluid bed and rotating shell processors, etc.

j. *Membrane processing systems*: cross-flow filters, ultrafiltration, nanofiltration, reverse osmosis, and pervaporation

k. *Fermentors and auxiliary equipment*: stirred tanks with special cooling and steam sterilization provisions, designed for ease of sterilization and maintenance of sterility, gas sparging, higher than usual power inputs through agitation, air sterilizers, feed tanks of various sizes, liquid sterilizers (16)

l. *Portable equipment cleaning modules, clean-in-place provisions in many vessels, fixed cleaning stations*

Process control capabilities

The bulk drug pilot plant must execute unusually varied processing with the requisite degree of control over the process variables, as well as have extraordinary means for data capture on-line (directly from sensors or analyzers in the equipment) and off-line (from samples tested in the laboratory and by derivation from raw data; e.g., the supersaturation profile of a batch crystallization, the performance of a fermentation cell

Table 3 Physical Plant of the Comprehensive Pilot Plant for Bulk Drug Processing (*Continued*)

mass from off-gas data analysis, the changes in the agitation requirements through the course of a reaction or fermentation, the heat balance across a condenser, etc.). Indeed, pilot plant work provides the opportunity to gather engineering data on scaled-up process performance during development, thus facilitating the better process design decisions. Often enough, the scale-up data drive the development of the process in a different direction as well.

Accordingly, most vessels and other equipment are provided with the appropriate sensors (temperature, pressure, level, pH, rotational speed, flow rate, weight, conductivity, etc.) and on-line analyzers as required. Some of the sensors may be used for local read-out (the value of the process variable may be read at the location of the equipment) and as inputs to a process control system elsewhere (a control room where, among other things, the inputs are converted by the controllers to outputs to valves, switches and actuators in the field). The rest of the sensors may be used for local read-out and control (the control device is also at the location of the equipment), but may also share the read-out with the control room. The option of operating through local control exclusively, while still available in principle, is rather unlikely to be found, as even operating environments of modest scope have mostly strived for some degree of remote control.

A sensor for the process control loop consists of these principal components:

- A sensor for the process variable at the appropriate point in the process stream and equipment, e.g., a thermocouple that generates a voltage as a function of temperature. The voltage is received by a transmitter (usually located at the equipment) that converts the voltage to a signal recognizable by the next device, e.g., a current in the 4–10 mA range according to a preset calibration of temperature to voltage to current.
- A controller (or control device) that receives the input signal from the sensor and transmitter, and compares the value of the process variable with a target value (the set point) and sends out an output signal to adjust the variable as needed.
- An actuator (a valve, a rotational speed drive, etc.) that, in response to the output signal from the controller, seeks to adjust the process variable. For example, the actuator may be a control valve that allows more steam to pass through and thus increase the temperature of the process materials in the equipment.

This said, remote process control can be variously implemented, taking advantage of the many scopes of control systems that are available. Ambitiously, a pilot plant processing area will have most of its process and

(*Continued*)

Table 3   Physical Plant of the Comprehensive Pilot Plant for Bulk Drug Processing (*Continued*)

infrastructure variables controlled remotely from a central location, using digital control devices connected to, with many governed by, a substantial computer system. The latter monitors selected process variables and controls many of those, relying on schemes that range from individual loops, for which the set point is entered at will, to complete schemes (and their algorithms) that sequence and control the processing events from a master set of instructions (often referred by the unfortunate term "recipe") or maintain the readiness of the infrastructure (HVAC and utilities). Most ambitiously, as well as unwisely, the designers of the control system may reach out and link the system that controls the processing events with extraneous systems, such as those that manage in-process and QC data or those that manage materials inventories and procurement. Such excessive connectivity is not needed for process control or process data gathering, but greatly increases the burden of the inevitable validation task. While perhaps of value in a large manufacturing environments, such linkages and data sharing aid the pilot plant task marginally at the cost of greater validation and system maintenance efforts.

Provisions for ascertaining the identity, lot number, and release status of materials, such as bar coding and the like, should not be viewed as excessive, but should be well isolated from the direction and execution of process tasks. Indeed, the objective of the minimalist approach is to reduce to the minimum those connectivities that are superfluous to the basic task of a pilot plant: simultaneously prepare material and develop the chemical process.

Indeed, a very adequate and prudent approach is to delegate to local microprocessors (or programmable logic controllers) the lesser control tasks, for which modern equipment comes with fully developed and validated process control packages; e.g., automatic centrifuges for filtration, manipulation of heating and cooling fluids in vessel jacket services, HVAC management systems, etc. In this lean approach, and *to the extent possible*, the supervisory process control system (at the top of the hierarchy) simply triggers the actions of subordinate systems and, during the period of action by the latter, it may monitor the appropriate process variables but does not control them. At the processing level, the subordinate systems do all the manipulations under the benevolent gaze of the top system and fade out of the scheme once the task is done; they do not link with the supervisory system except to acknowledge the instruction to start or to indicate its completion. Also in this approach, the supervisory system operates at arms length with any extraneous systems involved with materials or laboratory data management. The objective, of course, is to severely limit the range of unintended consequences associated with large sets of computer code controlling multiple tasks and manipulating large amounts of data. As the regulatory expectations on the integrity of control and other software-based systems

(Continued)

Table 3 Physical Plant of the Comprehensive Pilot Plant for Bulk Drug Processing (*Continued*)

---

approach the fastidious, simplicity in the design of schemes for process control and data management becomes compelling and, hopefully, there will be a persuasive minimalist among those making such design choices.

The qualifier "to the extent possible" recognizes the fact that certain tasks are beyond the ability of the individual equipment control package, as well as the desirability of setting up *at will* specific control loops and their set points, the sequence of events, alarms, etc.

Striving for such simplicity is not to be confused with the now ascendant SOP-based method of processing (standard operating procedure). In the latter, the execution of the processing can approach a veritable daisy chain of SOPs that minutely dice the overall task. In the extreme, the so-called manufacturing document (or whatever term might be used) becomes little more than a log, offering no perspective on the processing. Worst of all, operating personnel understand what they are doing dimly at best, and mining the document for troubleshooting or assessment information is tedious and often unproductive.

Portable control modules may also be used; these consist of recorders and controllers to create local loops with the appropriate sensors and valves or actuators. Specialized analyzers can be used on ad hoc or permanent basis for on-line analysis of specific process variables, often through a sampling loop.

A laboratory for in-process control testing must be equipped and staffed well, and managed to be responsive and convenient to the pilot plant processing areas. In the pilot plant environment, there is a large data gathering component that often results in large loads of in-process testing relative to manufacturing operations.

Finally, and still in the process control subject, it is important to provide the technical and supervisory staff with suitable and convenient office space and related work areas. In all chemical processing, regardless of the control scheme being used, the eyeball contact with the process operations is important. This is the case, most of all, for the bulk drug pilot plant, as a great fraction of the activities are being done for the first time and repetitive processing is so infrequent.

Traninng

Training facilities and unrelenting training are an indispensable part of a bulk drug pilot plant operation, aimed at safe processing, reliable preparation of bulk drug and effective use of the opportunity to scale up the process and gather the desired know-how. Given the experimental nature of the pilot environment, those that run it need to be extraordinarily sensitive to the process to be run. In that respect, SOPs have a rather limited value, as only the most elemental actions are repetitive; today's batch incorporates significant differences from yesterday's and will be different from tomorrow's, even if performing the

---

(Continued)

Table 3 Physical Plant of the Comprehensive Pilot Plant for Bulk Drug Processing (*Continued*)

same basic process. Thus, training must be *based* on the fundamentals, which can be effectively presented to all involved.

The increased use of SOPs, driven by a regulatory preference and the seemingly paramount objective of consistency, can, in the extreme, dice the operating instructions so minutely (approaching a daisy chain of SOPs) so as make it very difficult for the operating personnel *at various levels* to fully appreciate the scope of the overall task and the linkages between its different parts. This, coupled with the prevalent use of great detail in the operating instructions, has led to documents that are unwieldy, replete with discontinuities and very hard to use as training tools or for troubleshooting or retrospective data mining.

While undoing this state of affairs may, alas, not be possible, it is indeed possible, and most advisable, to preface the formal operating instructions document with a brief outline of the process and its procedure, written in clear prose (not instructions in the imperative mood) and accompanied by a flow diagram of the procedure in the context of the designated equipment, as well as including some brief discussion of the objectives of the work to follow. For example, a statement such as "set up RE-302 for distillation under total reflux prior to the application of steam its jacket," an instruction that in today's documents may take a number of subinstructions and no less than a page, conveys quite clearly the operational intent and was, at one time, quite sufficient for a skilled chemical operator. A preview of such preface by all levels of operating personnel has multiple advantages, including alleviating the insidious effect of the disrespect for the operating people implicit in the prevailing mode of operating instructions.

Besides the physical plant there are, of course, other capabilities and attributes for a successful bulk drug pilot plant environment. To wit:

1. A skilled team of chemical engineers and chemists to operate the facility, with the depth to operate as a process development cadre, capable of addressing the full range of technical needs of scale-up and process design work as well as day-to-day operation.

2. Close collaboration and ready access to bench development chemists, microbiologists, and engineers who have project (drug candidate) responsibilities in the pilot plant as well.

Roof with equipment

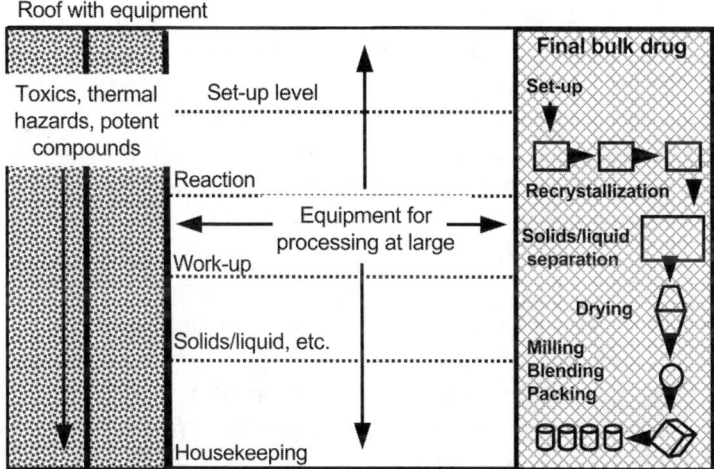

Figure 14 The multipurpose, once-through gravity plant for bulk drug processing.

3. A skilled team of chemical operators that are trained unceasingly in the fundamentals as well as in all that is new by way of equipment, procedures, policies and applicable regulations.

4. A support laboratory for close and responsive in-process, troubleshooting, and QA/QC support to the pilot plant operation.

5. Ready access to the analytical R&D function of the bulk process development area.

6. For fermentation processes, the appropriate lab capabilities as per 4. and 5. need to exist. These differ markedly from those of chemical synthesis (16).

7. A skilled materials management function with the appropriate tools for materials tracking, documentation, and security.

8. A dedicated maintenance and minor installation team with adequate workshop and stores.

9. A skilled clerical support function, well trained on the regulatory obligations of the pilot plant operation and the requisite tools.

10. Internal skills in the environmental engineering field and the applicable regulatory milieu for the pilot plant facility, as well as ready access to the appropriate site or corporate functions.
11. Internal skills in the operational safety and industrial hygiene fields, as well as ready access to the appropriate site or corporate functions.
12. Management systems and a managerial tone that foster, and insist on safe and responsible operation, strict maintenance of the physical plant, continuous training and strict regulatory compliance as called for by the developmental activity.
13. A management that foster and maintains a pilot plant organization as a vibrant, engaged and highly skilled component of a broadly based process development function in the bulk drug business. An indispensable obligation of the pilot plant management is to ensure that no one forgets, under the pressure of serious operational and regulatory demands, that the pilot plant is an experimental environment with a major responsibility in the creation of the process body of knowledge.

Finally, and as indicated in Table 3, the pilot plant processing equipment needs to be set up and tailored to facile data gathering, well beyond the usual process variables measurements—extraordinary sampling ports and devices, nozzles, and flow loops set aside for the insertion of infrequently used sensors, recirculating sample loop modules, flow loops, etc. Modern pilot plants are usually well provided with process control systems that monitor, control, and have responsive sequencing capabilities. Such systems are very advantageous in what is, after all, an experimental environment. One needs to be alert, however, to the possibility of making the process control system too rigid in its operating procedures, and thus discourage the enterprising experimentalist or data gatherer. Finally, the operating style of such a modern facility should still encourage old-fashioned eyeball contact with the process on the plant floor.

In summary, the bulk drug pilot plant is a critical mass of skills and capabilities within the larger critical mass of the bulk process development function. Obviously, the size and scope of the organization matters a great deal—bulk drug pilot plants with capital replacement values of over a billion dollars and operating budgets well over US $100 million per year exist. Nevertheless, and although difficult, pilot plants of lesser scope and ambition can be created *provided* the requisite skills and management systems are assembled cohesively and maintained well.

D. New Processing Technologies

The process development environment is optimal for the evaluation of new technologies and methods for bulk drug processing, as all the necessary elements exist and are well poised for the acquisition of new experience:

a. the aggregate of discipline skills;
b. the interdisciplinary critical mass;
c. the experimental capabilities at bench, kilo lab and pilot plant scales;
d. the working interfaces with process design and manufacturing.

Yet, there is an element of risk that, albeit of a different character, may be seen as comparable to that encountered when evaluating new technologies in manufacturing. Whereas the latter is burdened with rigid regulatory constraints that may ultimately quarantine or preclude the sale of product made under test conditions, the pilot plant is comparably constrained; not because of the regulations, but because of the risk to the supply of material to the drug development programs. This risk is, for all practical purposes, regarded just as large as, or larger than, lost manufacturing output, as there is a potential impact on drug development timeline.

Alas, the seemingly obvious solution of evaluating new technologies on a parallel track does not work well enough. At some point, the new technology or method needs to be reduced to practice at scale and the perceived risk arises— compromising the yield or quality of material made under the test conditions or usurping preparative capacity for non-preparative purposes. In a vigorous drug development program, that capacity (technical personnel as well as equipment) is usually fully allocated. Furthermore, R&D management at large has no sympathy for such distractions, a utilitarian outlook that could be well justified.

The less obvious solution, however, is to evaluate the new technology in stages, not unlike a new process variation arising from the development work, such as a different starting material, solvent, or catalyst, an improved purification or a faster and more reliable drying procedure for the bulk drug. These latter changes are routinely introduced to the preparative work in a deliberate manner, but with the relative procedural ease that characterizes the pilot plant environment. R&D management knows, perhaps deep in its subconscious, that the bulk process development function merges the preparative work with constant scaling up of new methods, and that a very large fraction of preparative work is also experimental. With discretion and with an extra measure of deliberation, new technologies can be similarly evaluated at no greater risk.

In that regard, the more experimental aspects of the pilot plant that have been described under III.B are very well suited. In addition, getting the moral, as well as other support from the manufacturing organization adds to the impetus and justification of the apparent distraction of bulk development resources. This is even more important to the technology stance of bulk drug manufacturing under the current and foreseeable regulatory environment that unwittingly discourages innovation into pharmaceutical manufacturing.

E. Beyond the Pilot Plant

As the *consolidation stage* comes to a close—slower pace of preparative work and having provided the body of knowledge contribution to the assembly of the dossier—the bulk development team shifts its focus to the technology transfer to manufacturing. Although its participation in the preliminaries started, or should have started, at the early *development stage* and continued, increasingly, through the *consolidation stage*, now the time has come to demonstrate the bulk drug process performance in the first manufacturing plant or plants, the latter in the event of multiple sites of first manufacture. In such cases, the bulk process is generally operated in one plant through the bulk drug, but a slip stream of penultimate compound (the final intermediate) or the final compound in unfinished form is shipped to another site for final processing to the bulk drug.

There is considerable material to cover on what happens beyond the pilot plant, and such is the subject of Chapter 3.

IV. THE PHYSICOCHEMICAL ATTRIBUTES OF THE BULK DRUG

As one of the three basic tasks of bulk drug process development, defining and achieving the physicochemical attributes of the bulk drug is pursued throughout the development cycle. Unavoidably, this effort trails that of the chemical or fermentation process, since its target comes from the dosage form development effort.

The difficulty of the dosage form task cannot be underestimated. Its need for making judgments with partial data actually exceeds that of the bulk development task, as the crucial feedback on the bioavailability and stability of its developmental materials cannot be obtained rapidly, not unlike the feedback that the bulk development team needs as to the suitability of the its bulk drug for the dosage form purposes. Figure 15 is an attempt to depict the scope of dosage form development.

> Not to be neglected is the packaging development placed directly downstream from dosage form development. Sometimes complicated by the fact that the primary package (i.e., that in direct contact with the drug product) may also serve as a drug delivery device (e.g., syringes, eye drop dispensers, intravenous bags) as well as by the issues of interaction between the dosage form and the package's material or the long-term stability of the drug product within the particular package chosen, the dosage form development function is additionally

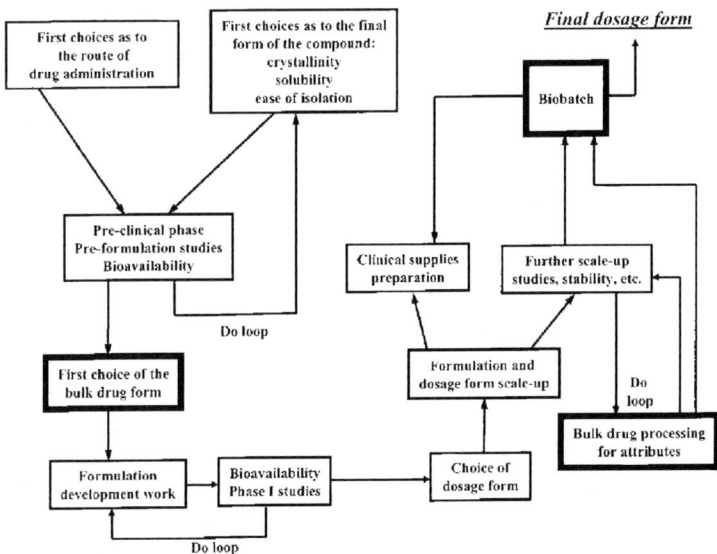

Figure 15 The scope of the dosage form development task. The notation "Do loop" refers to the iterative process by analogy to the Fortran language shorthand.

> buffeted by marketing issues that range from the serious (acceptability by the patient, for example) to the seemingly frivolous (for example, the marketer's insistence on a distinct tablet shape that, although harder to manufacture, lacks any apparent redeeming value).

Albeit hampered by a traditional disciplinary divide between pharmacy and the disciplines of bulk process development, the bulk/dosage development interaction needs to start early and intensely. Not only is the task difficult for the reasons just stated but also there is considerable scope to getting to a firm definition of what is needed. Figure 16 lists the physicochemical attributes of bulk drugs that must be controlled in the bulk drug process, either directly, such as particle size, or indirectly, such as solid surface area or hygroscopicity.

Those attributes are, of course, set by the very last processing steps of the bulk drugs:

    a.   the last synthesis step (or the last purification step if a fermentation/extraction process);

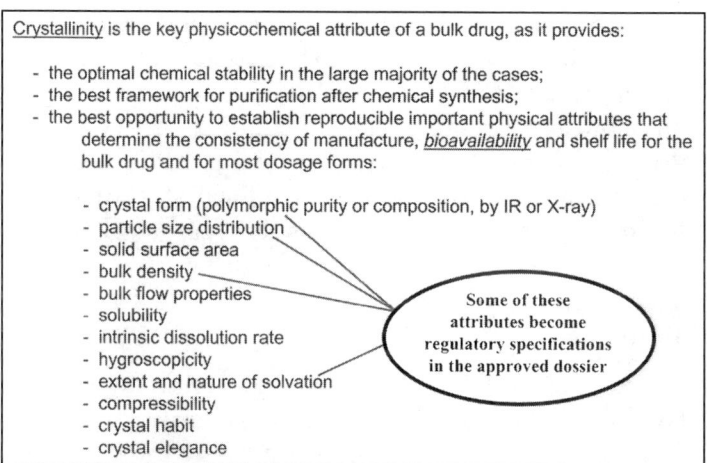

Figure 16 The physicochemical attributes of a bulk drug.

b. the subsequent isolation (usually by crystallization), filtration, and drying;

c. the final solids finishing (size reduction, classification, blending, and packaging).

Due to the significance of the physicochemical attributes of the bulk drug to its bulk and dosage form stabilities, as well as to the dosage form performance (mostly bioavailability), it has become increasingly frequent to add a recrystallization after the isolation of the final chemical compound and thus generate the *bulk drug*. Although such additional processing is expensive (its yield loss is incurred with the costliest compound), there are significant advantages to consider:

a. The final compound isolation is relieved form the dual burden of simultaneously achieving all the chemical purity attributes *and* all the physicochemical attributes.

b. The discontinuity makes the final bulk drug less subject to upstream variations from the more complex synthesis/purification/isolation step.

c. The final recrystallization can provide an additional degree of purification that may be reserved as insurance or be part of achieving the final purity.

d. The final recrystallization can be developed with a sharp focus on *consistent* attributes such as polymorphic content, crystal habit, particle size distribution, surface area, and bulk density. These attributes also define hygroscopicity and are important factors on bulk and dosage form stability.

e. Regulators are very fond of such recrystallizations (for the above reasons), which can be used to more persuasively present upstream process changes for approval.

The addition of such final recrystallization is depicted in Fig. 17. Given the fact that practically all bulk drugs are crystalline for reasons of processing soundness, purity, stability, and consistent physicochemical attributes, obtaining registration of a bulk drug in any other physical state usually requires a compelling reason.

As practitioners discover (or should soon discover), crystallization skills are paramount among the skills set of the bulk process development (and manufacturing) function.

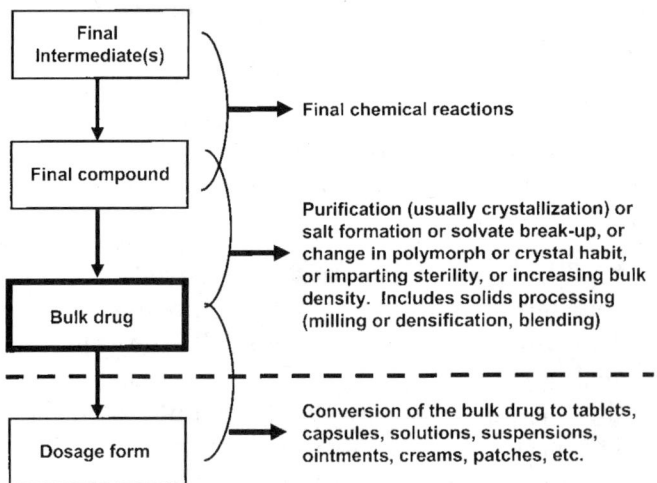

Figure 17 The final stages of bulk drug processing. A final crystallization is often inserted largely, if not strictly, for greater control over the physicochemical attributes of the bulk drug.

Such skills should be nurtured, perhaps even lavished, as well as complemented with a comparable physical chemistry capability in the analytical R&D function.

This seems a good point at which define the analytical R&D function as far more than the guardian of quality during the preparative work, or the highly skilled developer of assay and related methods, or the arbiter of regulatory issues within the bulk process development. For successful bulk drug process development, the analytical R&D function must be an integral part of the *process team*: elucidatior, troubleshooter, contributor to the solution of process problems, and intimate partner throughout. Any lesser involvement in the process task or a lesser aggregate of skills is a strategic disadvantage in drug development.

Finally, the bulk process and dosage form functions need to collaborate earnestly at the earliest and, if needed, be brought together under irresistible force to overcome the traditional disciplinary gap. For example, the crucial decision on which bulk drug form is to go forward (the sodium salt? the maleate? the dihydrate?) is best made when the bulk process team participates and is able to contribute its resourcefulness, lest the dosage form team abandon the better bulk drug form because they do not know that its difficult attributes can be managed or fully overcome upstream. It is there that those attributes are set through the actual process in the final reaction step and in the finishing steps that follow to afford the final bulk drug.

V. THE PROCESS BODY OF KNOWLEDGE

In developing the process for a bulk drug, the need to gather and properly organize a process body of knowledge is compelling:

1. The dossier requires that the process foundation—chemistry, engineering, scale-up, bulk drug chemical and physicochemical attributes, environmental impact, process controls, preparative and developmental history, and bulk/dosage issues—be readily available and well organized for the assembly of the various individual submissions. Increas-

ingly, these submissions need to provide a level of *bulk drug process validation*; namely, a persuasive case that, on the basis of the actual performance during the preparative and developmental effort, the process is capable of disciplined manufacture and thus result in bulk drug output that is consistently safe and efficacious.

Additionally, the dossier should have the documentary basis for being able to include in the submissions a similarly persuasive case that the proposed manufacturing plans are generally sound. To wit, that the intended manufacturing milieu will do justice to the process needs as identified during its development. Such bridge *documentation*, while not relieving the technology transfer team from the burdens of whatever inspection the manufacturing plant may face, greatly facilitates the task of preparing for such inspection and for dealing with it if it occurs. Just as importantly, the said *bridge documentation* also provides the regulatory reviewer with sufficient background to answer a number of probable questions and thus avoid their being asked during the review cycle.

> It is rather easy for practitioners, when faced with the above tasks, to get thoroughly lost in the detailed "how to" without fully grasping the scope and objectives of the dossier on the process, which have been defined above. As the literature on "how to" grows and seminars, workshops, and guidance documents proliferate, the practitioner should first seek a clear understanding of what it is that the submission review and the plant inspection *basically* seek to accomplish. The above paragraphs are an attempt to clearly define just that, and additional background material, still at the usefully general level, is suggested (18,19).

2.   The technology transfer to manufacturing demands that the process be well documented. Most important, and as a coalescence of the process know-how, a *comprehensive process document*, written for the specific purpose of imparting knowledge, is a requisite for the tasks of:

a.   process design;

b.   project engineering design and construction;

c.   procurement of materials;

d.   preparation of start-up plans and operating procedures;

e.   transfer of the in-process and QC analytical methods;

f.   assessment of the process safety issues in the specific context of the plant: operational safety, industrial hygiene, thermochemical, and environmental safety;

g.   assembly (and timely approval) of environmental and other regulatory permits;

h.   definition of the process start-up targets of yield, capacity, waste loads, etc.;

i.   dealing with assorted other matters, such as those arising from the plant's insurance, etc.

It is, of course, unacceptable to bind together all manner of development reports and send them over with a cover memorandum (part of what is aptly known as "over the fence" technology transfer). Attachments are very important, but a well-edited document that is rich in content and aimed at guiding the downstream practitioners is the indispensable first vehicle for the transfer of the know-how. Not even the most thorough collaboration between development and manufacturing can completely remedy the lack of the above comprehensive process document intended for imparting knowledge.

There is no attempt here to gloss over the extraordinary effort and discipline required to turn out such documentation on a timely basis. The *consolidation stage* is intense, the gentle slope of the know-how curve notwithstanding. Yet, the quality of the technology transfer—on both the short and the long term—is very much enhanced if such a document is available soon enough.

Conversely, and as indicated in Figs. 11 and 13, the joint effort with the operating organization (planning and process design) unavoidably overlaps the actual development of the process, sometimes to a great extent; e.g., if an early decision is made to build a new plant or substantially alter an existing plant, the downstream work cannot await a sufficient definition of the process, and information needs to flow as the process is developed. This is a very demanding task for all involved that benefits from considerable practice, significant skills of process design on both sides and from a spirit of collaboration, preferably steeped in previous joint successes. Unlike other chemical processing activities, new bulk drugs are exceedingly driven by the "time to market" imperative and organizations that can significantly overlap process development with manufacturing readiness work have a strategic advantage. However, having such skills and practices do not relieve the development team from the duty of comprehensively documenting the process at the earliest reasonable time.

3. After successful technology transfer, which must, of course, be well documented also, the original process body of knowledge serves as the foundation for management of the change control system, for training of new manufacturing personnel and as the basis for sound process improvement work. Indeed, significant second-generation processes are most often based on approaches suggested and partially elaborated during the original development.

4. Although the interaction with suppliers and contract manufacturers will be discussed more amply under VII, it is often that process information needs to be transmitted to outsiders, including prospective licensees of the drug candidate. Indeed, this happens most likely during the developmental phase as help is sought in the preparation of intermediates, the bulk drug itself or in further development of the process or an alternative route for which the outside collaborator may be better positioned.

It is in such instances that having a system for *continuing* process documentation pays off in the rapid satisfaction of needs that may arise unexpectedly. Ideally, the material should exist in organized form so as to permit knowledgeable technologists to assemble and edit a preliminary process documentation package in a matter of a few days and a full package in, say, 2 weeks. Of course, the transmittal of internal documents "as is" is fraught with the risk of undue disclosure and it is best to transmit documents *assembled and edited for the specific purpose at hand,* a task hardly feasible if the material does not exists or exists disjointed or incomplete.

> The obvious need to avoid undue disclosure of internal issues and business methods, names, distribution lists and the like, as well as to avoid transmitting information extraneous or strictly tangential to the technical matter at hand, must not be confused with undue reticence. If others are expected to properly implement in-house process know-how or to use it as the basis for an activity to be done on our behalf or as part of an agreement or license, the disclosure of the technical information must be no less than sufficient: what works and what does not work, the technical rationale for the prior decisions, our best understanding of the process issues and sufficient detail of methods, process design calculations, and data. In particular, data on thermochemical safety, industrial hygiene, and environmental profile need to be fully disclosed.

5. Developing the appropriate intellectual property is also greatly facilitated by the continuing process documentation system being advocated herein. Laboratory notebooks and pilot plant log or batch sheets, while useful for assigning dates of reduction to practice, compositions of matter, procedural details and for the identification of inventors, are generally inadequate sources of cohesive process information and history.

6. Finally, there is the organizational objective of fostering a professional climate for the process technologists to thrive. The rigors and satisfactions of authorship of scientific and technical documents arising from one's own work are not to be underestimated; they contribute greatly to the individuals and to the organization as a whole, even if intended for internal publication only.

> The ready access to powerful computers has created an environment in which databases and templates or excessively formatted documents are quite seductive as a seemingly easier substitute to a system based on documents composed in clear, informative, and persuasive prose. Thus, in such tempting systems, the process know-how can be thinly dispersed over an alphabet soup

of spreadsheets and form-like documents that, inevitably, lack the full benefits of reflection and perspective from an author (or authors) with a process story to tell or a point of view to present as to how to implement a process. Such temptation should be resisted, as extracting useful and applicable process knowledge from the former environment is not possible without a substantial effort of retrospective composition that would have been better applied to the creation of true process documents.

Scientists and engineers, usually handicapped as writers by the focus of their academic training and by misconception as to the scope of technical writing, are destined to further disadvantage if nudged by managerial convenience or by conformity into documenting their work as if filling blanks in a form, or seeing the process body of knowledge as an array of suitably filled pigeonholes.

The effort in setting down and organizing the process body of knowledge should not trail the acquisition of the raw inputs, as tardy heroic efforts to properly document accumulated knowledge are invariably not as good as the task deserves. Figure 18 outlines the body of knowledge task on the applied effort vs. know-how plane.

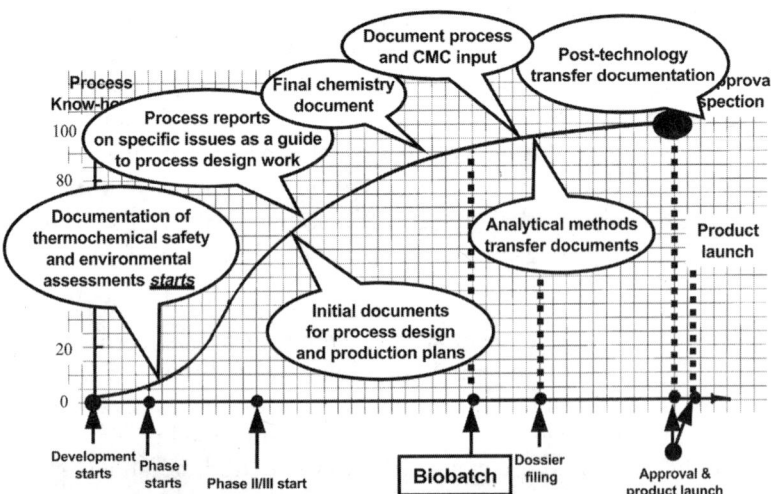

Figure 18 The process body of knowledge in the know-how vs. applied effort plane.

Finally, Table 4 offers an annotated template that, if followed with sufficient discipline, carries out the various missions of the bulk drug process body of knowledge. Of particular value are the milestone reports and the specific issues reports, as they permit achieving depth and focus, while nurturing vigorous authorship by the process technologists. Such documents are invaluable as part of the comprehensive process documentation, as well as excellent raw material for the bridge documentation of the dossier. Their elaboration into external publications is usually a much lesser effort than starting from raw data and status reports. The following examples illustrate the proposed reports on milestone events and specific processing issues. Note that the titles of these reports have been composed by this author as fictitious from literature sources or approximate from his own experience with actual bulk drug process projects:

Table 4 The Scope of the Process Body of Knowledge and Its Applications

Document	Frequency	Target (1)			
		R&D information	R&D archival	Process design effort	Manufacturing organization
Status reports from each discipline & function	Quarterly				
Kilo lab experience	Milestone				
Pilot Plant - Preparative experience	Milestone				Summaries
Chemistry - Specific reports	Milestone				
Microbiology - Specific reports	Milestone				
Engineering - Specific reports	Milestone				
Analytical - Specific reports	Milestone				
Thermochemical safety reports	Milestone				
Environmental assessments	Milestone				
Industrial Hygiene matters	Milestone				
Process economics	Milestone			Summaries	
Chemistry - Comprehensive background	Pre-filing				
Microbiology - Comprehensive background	Pre-filing				
Engineering - Comprehensive background	Pre-filing				
Analytical - Comprehensive background	Pre-filing				
Environmental assessment - Final	Pre-filing				
Stability studies	Milestone				
(2) for technology transfer purposes	Milestone				
Biobatch report	Milestone			Summary	
Bulk drug attributes report - Final	Milestone				
Analytical methods - Final	Pre-filing				
Stability report - Final	Pre-filing				
Process demonstration document (3)	Milestone				

Notes:
(1) Darker shade denotes the principal targets, whereas the lighter shades denotes other recipients of the document.
(2) Includes thermochemical safety and environmental. Attaches chemistry and microbiology documents as needed.
(3) Joint with the Manufacturing organization.
(4) Milestone denotes that the document is generated upon certain outcomes or decision points being reached. Pre-filing indicates that document is issued *specifically* for assembly of the dossier.

ICI 194008. The benzaldehyde imine route to the amine tosylate precursor. Bench development and readiness for scale-up (5, page 22)

MK-787 via the ADC-6 chiral route. Results and experience from the first large-scale pilot campaign (C. B. Rosas, personal communication, 2003).

Efrotomycin. Whole broth extraction in the mixer settler and in the centrifugally aided extractor. Results and recommendations for process design at the Stonewall plant (C. B. Rosas, personal communication, 2003).

MK-401. Early environmental assessment of alternatives for the trichloro precursor (C. B. Rosas, personal communication, 2003).

MK-421. Large-scale synthesis of AlaPro in a continuous flow system. Process design and results obtained at the large pilot scale (20).

Diazomethane. Pilot scale generation by continuous reaction and scale-up criteria for the commercial scale (21).

LY228729. Kornfeld ketone route as the selection for scaled-up development (22).

These two kinds of reports, when added to well-designed status reports that issue regularly and not too frequently, provide the basis for a repository of a well-organized body of knowledge that can be used for the various objectives previously defined. Indeed, such reports are the core of the body of knowledge, as they gather, coalesce, and make cohesive for application the great deal of data and experience gathered during all aspects of process development.

Other aspects of the system in Table 4 are:

1. All the reports and documents listed originate in the bulk drug process development area, which embraces the disciplines and function shown in Fig. 12, and is part of R&D at large.
2. All process documents intended for the process design effort originate from the engineering discipline (chemical and biochemical), which embraces

the thermochemical and environmental safety functions within R&D. Material from the other disciplines is attached as required.

3. All process documents intended for the dossier assembly are generated *specifically* for that purpose, usually through a CMC function (as per the chemistry, manufacturing, and control component of the NDA). Such function is within the bulk drug process development area and not within regulatory affairs. This latter function should not use other process documents for its purpose of assembling the dossier or attempt to edit regular process documents on its own.

4. The biobatch, although an event taking place in the dosage form development area (and documented accordingly), will usually generate the need to document the process and related history of the bulk drug inputs used.

5. The process document is generated upon completion of the technology transfer to first manufacture and is coauthored jointly by the bulk drug development area and the recipient manufacturing organizations. Chapter 3 discusses this and all other aspects of the technology transfer in some detail.

VI. PROCESSING RESPONSIBILITY IN BULK DRUG PROCESS DEVELOPMENT

All chemical processing, whether on a large or a small scale, whether for high value chemicals or commodities, or for bulk drugs, textile polymers, petrochemicals or household products, carries a risk to those that work in the industry, to people around the manufacturing sites and beyond, and to the environment: locally, beyond the locality, and at large. Indeed, the risk comes about from multiple directions:

1. The hazards created by the chemistry itself: (a) intended and unintended energy releases, and (b) the various hazards of handling the materials involved. Bulk drugs often

present a peculiar hazard, i.e., the relatively high potency of the desired biological activities, as well as the collateral activities of the intermediate compounds and, of course, of the drugs themselves.

2. The specific manner in which the chemical processes is implemented at scale. Most risks in chemical processing are a function of the process design, the equipment design, and the operating procedures used to manufacture the products. In other words, *the same inherent hazard can be implemented at various levels of risk depending on the specifics of implementation*, and often enough details matter.

Hazard—a source of danger, of possible injury or loss.
Risk—The probability of suffering a given loss or injury from a hazard.

3. The local context in which the manufacturing process is implemented. First, there are factors, such as the proximity to populated areas, the direct impact on sensitive receiving bodies of water or other valuable habitats, or a less apparent impact on remote parts of the environment at large. Then, as a lesser subset of those risks are the various statutory and regulatory constraints that create liability potentials or that may impede timely manufacture if not properly addressed.

It is one of the prime responsibilities of the bulk drug process development organization to seek processes of acceptable levels of risk in both the chemistry and its engineering, and to participate in the process design and manufacturing plans to see to it that their implementation risk is sufficiently low. Of all aspects of technology transfer, none demands more in terms of the development team thrusting itself downstream and seeking the closest collaboration with the manufacturing organization. Clearly, the greatest opportunity for success exists at the developmental stage of the R&D process work, when the process is conceived and developed; engineering low risk into the implementation of a hazardous process is always the second choice for the bulk drug process development team and the collaborating process design function. For example, much safer process alternatives seemed to exist for the process that led to the 1974 catastrophe in Bhopal, India (24); one called for a different chemical route and the other for a different process design of the original chemistry.

Incidents such as Seveso, Italy, 1976 (24,25) and Bhopal illustrate the potential for catastrophic events from aberrant chemical processing and design, sloppy operating practices and incomplete knowledge about probable unintended events and consequences.

Generally useful practices in this aspect of bulk drug process development are:

- Early assessment to guide the process conception and choices. This implies availability within the process development organization of, or facile access to, laboratory capabilities to evaluate thermochemical and environmental hazards. The evaluation of industrial hygiene hazards is aimed at the protection of personnel, and is facilitated by the availability or access to adequate toxicology resources such as those generally available to a research drug firm. This industrial hygiene context, however, differs substantially from that of evaluating the risk to patients taking the drugs, and a different subset of skills and methods applies.
- Continuing assessment as the process develops, including a vigorous interaction with the process design function and the manufacturing organization. For example, issues such as the choice of manufacturing site, which influences the risk, cannot be settled by the process development team alone, nor can they be properly settled without the hazards assessed during development.
- Reasonably early decision on the in-house vs. outside manufacturing choices, as the latter requires technology transfer and due diligence work, as well as the inevitably longer cycle for reaching the necessary technical and business agreements (more on this under VII below).

A. Thermochemical Process Safety

Most chemical processing operations have energy exchanges between process streams and the surroundings; process streams are heated or cooled for various purposes and such exchanges need to be safe. Heated streams must not exceed limits that generate undue pressures or undesired chemical events, whereas cooled streams must not freeze and interrupt process flows, or hamper a desired chemical reaction and accumulate unstable intermediates.

> A distinction needs to be made between limits observed to maintain process performance and limits observed to avoid a hazardous operating condition. The above paragraph refers, of course, to the latter limits, as depicted in Fig. 19 using the safe processing envelope concept.

The thermochemical safety of chemical processing deals with the safe handling of the energy released from chemical reactions and with the prevention of unwanted releases of energy. Chemical reactants may, when converted to products, result in the transformation of chemical energy into heat, and during such exothermic reactions the heat release needs to be safely managed. In addition, chemical process streams may reach abnormal conditions that cause unintended exothermic reactions, with the attendant formation of unintended byproducts and release of energy.

Hence, the objectives of thermochemical process safety as a distinct principal component of processing safety at large:

- Identify all *intended* energy releases and determine their magnitude, rates, and byproduct releases, such as gas evolution and their composition. These determinations need to be made over the appropriate range of process conditions.
- Identify *unintended* chemical events and energy releases for *reasonable* hypothetical situations (e.g., excess temperature by loss of coolant or runaway, excessive evaporation of solvent, interrupted reaction cycle, etc.) and assess their magnitude, probable rate, and consequences with respect to containment, gas

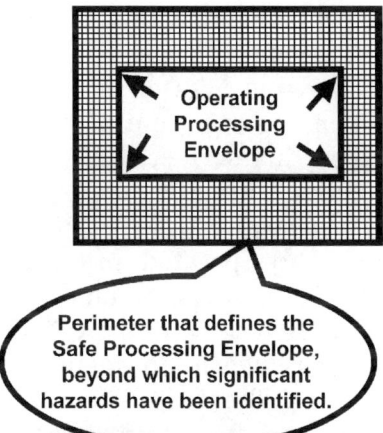

Figure 19 Processing limits for performance and for safety. Processing limits define the perimeter of the operating envelope that results in the range of desired process performance, whereas the safety limits define the safe processing envelope perimeter given the identified hazards that lie beyond. For example, a distillation is to be carried out at 90–100°C, whereas the high-temperature interlock that shuts off the steam is set at 125°C because a significant exotherm initiates at 160°C.

evolution and, when indicated, the composition and toxicity of the components of a plausible release.

- Identify and quantify the hazards of handling the process streams and materials with respect to shock sensitivity, flammability, explosiveness in air mixtures, dust/air explosiveness, etc.
- Seek process development solutions to avoid or reduce hazards. For example, one might seek an alternative reactant, a reaction medium that permits a lower reaction temperature or, in the ultimate, a different synthesis scheme for the conversions at hand.
- Provide process design solutions to those hazards that cannot be reasonably developed out of the process, thus reducing their risk to levels appropriately low for the operating context. For example, a hazardous

nitration reaction may be implemented in a reactor system that does not use aqueous coolants, or that is equipped with a suitable quenching vessel, or with a sufficient containment system, or using a continuous tubular reactor with large cooling surfaces and holding a small volume of reactive in-process materials. Similarly, a process or portions of a process with a hazard of explosion is preferably operated in a plant site that is distant from populated areas (vs. an otherwise more suitable plant site not as distant from populated areas); or a process with an identified hazard of aquatic toxicity in its untreated waste would not be operated in a plant site that normally discharges to an aquatic habitat. In both cases, one will take preventive measures to reduce the risk, but a risk differential will exist between the two plant sites.

From the above, and most importantly, *the practice of thermochemical safety far transcends the evaluation or the assessment of hazards.* It also demands that skillful solutions to the hazards be provided so as to eliminate them or reduce their risks as required. While the reader may view this statement as redundant or exceedingly tutorial, the fact is that a functional discontinuity between the assessment of thermochemical process hazards and the implementation of the process frequently exists, creating an ever present pitfall for the unwary, the sloppy, the overwhelmed, and the unqualified, and even organizations with the requisite critical mass of skills and well-documented procedures need to be vigilant to the gap. As in most other aspects of bulk drug process development, the utmost integration of process development and process design is the best approach to thermochemical process safety, organizational divides notwithstanding.

Additionally, the above reflects the fact that the same hazard (e.g., a reaction mass that can decompose explosively upon total loss of solvent) poses different risk according to the context of implementation. Thus, for the example just given of a major hazard, a process

design based on operator's attentiveness and simple process controls would entail a greater risk than a design
based on interlocking and redundant measures to prevent total loss of solvent as well as operator's attentiveness. To wit, the operational risk arising from a process
hazard is very much a function of the specifics of the
operational context.

There are, of course, many other aspects of processing
safety that are unrelated to, or overlap with, thermochemical
process safety. Among the overlapping, fire and explosion
hazards due to flammables handling stand out, whereas the
unrelated (e.g., falls, burns, asphyxiation in enclosed spaces,
static electricity, etc.) are generally addressed through the
aggregate of well-established measures of operational safety,
facility design, insurance policy expectations and applicable
industrial or building codes.

It seems best, even for the introductory scope of this
chapter, that before approaching a more specific discussion
of the fundamentals and the practice of thermochemical
process safety, the presentation of a broad perspective be
attempted. Hence Fig. 20, in which the field is viewed from
a sufficiently high vantage and that the reader is urged to
examine in earnest before going further.

Three key points arise from Fig. 20:

- The chemistry defines the overall scope of the hazards:
  the energy release potential of the reactants and other
  materials used, that of the reaction and process streams
  generated and the toxicity hazard that attends to all the
  compounds involved, whether inputs, intended, or generated by aberration. Accordingly, chemical acumen is
  utmost in the assessment and follow-up of the hazards
  defined by the structures at hand.
- Upon assessment, a broader set of skills is needed. Will
  the hazards be avoided altogether by a change in the
  chemistry or will its risk be sufficiently reduced by a
  process solution? Either approach requires engineering
  acumen to determine that a process solution is not
  advisable or probable, or to devise a suitable alternative.

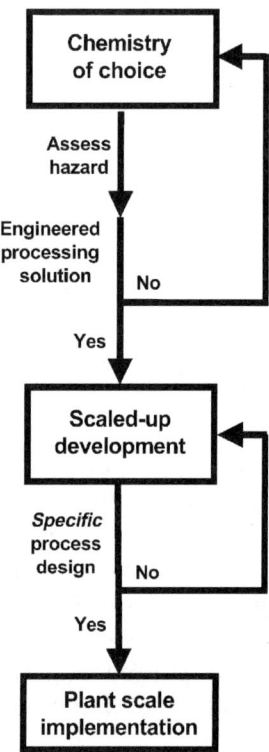

Figure 20 Thermochemical process safety in bulk drug process development.

- Finally, the implementation of whatever process is arrived at through development, and its indispensable process design collaboration, must go through further engineering analysis, by which all the applicable considerations must be pursued to the requisite level of detail: from the sizing of vessel relief and area explosion venting on the basis of thermochemical and related data to the evaluation of risk scenarios that will dictate the necessary margins of safety relative to overlapping safety, site specifics (e.g., weather precedents, proximity to people or valuable environments), applicable regulations, insurance policy expectations and all the way up to the probable perceptions in the neighboring communities.

Prior to the catastrophes in Seveso in 1976 (24,25) and in Bhopal in 1984 (24), these admonitions would have seemed unwarranted and even melodramatic, but not any more. Finally, seeking relief in the small scale of bulk drug chemical processing does not help, as both instances of chemical processing operations gone badly awry were of small scale.

1. Hazard Assessment and Methods in Thermochemical Process Safety

Thermochemical hazards are numerous and richly varied in kind, each requiring more than passing consideration and, if appropriate, an assessment by engineering design calculations, simulation, experimentation, or both. The task calls for experienced good judgment, as the possibilities are too numerous. For example, an organic synthesis of six distinct steps, with up to, say, 10 distinct intermediate structures generated, might also have a total of 40 different material inputs and process streams. Experimental assessment of each is a large burden that, invariably, can be greatly reduced by the said experienced good judgment.

Herein there is not, of course, the aim to comprehensively present this subject. Indeed, the literature is ample (not surprisingly, most was written after the 1976–1984 experiences), and the serious reader is earnestly referred to various references, preferably in the listed sequence (24,26–28). Clearly, this is not work for the dilettanti, but for professionals willing to invest in acquiring and applying focused know-how in a multidisciplinary environment– too much is at stake. Similarly, firms engaged in bulk drug processing cannot approach the work in half measures, or contract it out indiscriminately or unaware of the pitfalls of doing so.

Let us discuss another perspective, this time from a closer vantage; that of the thermochemical hazards assessment. Firstly, as indicated above, the structures at hand provide very useful leads as to what to expect. As a good rule of thumb, organic compounds that are rich in nitrogen, oxygen, or both are high on the list of reactivity and energy release structures, followed by some specific bonds and then by the less obvious cases that exist in organic synthesis, albeit less frequently (26, pp. 18–28; 27, pp. 28–52; 28, pp. 22–27). Once so alerted, the hazards assessor has a good number of techniques for estimation of heats of reaction, for rapid screening of exotherms and instabilities in materials, compounds and process streams, for accurate calorimetry work under close to actual process conditions and for very specific follow-up of hazardous conditions (27, pp. 1–28 and 52–88; 28, pp. 27–45). Indeed, the techniques are so numerous that care must be taken to walk the fine line between necessary and marginal testing, striving to reserve the more elaborate and exhaustive methods for the cases that merit them. For example, the thermal stability of process materials and streams can be pursued to great lengths (29) as required. Similarly, the subsequent hazards of vapor or gas release or toxicity of the released materials need to be pursued with similar acuity, as one may go too far as easily as not far enough. In this, the incisiveness of the screening effort makes the difference.

The well-executed hazards assessment meets its three basic objectives: (a) identifying and quantifying the heat effects of the intended chemistry, (b) identifying and quantifying, albeit not always as precisely, the thermochemical energy hazards from aberrant conditions, and (c) identifying and quantifying those hazards associated with the handling of process materials and their instabilities. Again, it is work with a great many nuances for which chemical, physicochemical, and engineering acumens are indispensable.

By way of vivid illustration of these assertions, one might consider the following instance, in which a labile nitrogen-rich compound was isolated as a water-moist powdery solid and dried under vacuum at ~50°C. These latter drying conditions had been set at ~50°C away from the rapid and large exothermic decomposition of the compound, found to initiate at ~100°C in the screening work. Additionally, the heating medium used in the drying step was limited to ~55°C and an ample vacuum capacity and a suitably low terminal pressure provided for the thorough removal of water. After months of processing at the ton scale, a process change was introduced in the isolation, substituting a mineral acid for another in the final acidification prior to filtration and washing. This seemingly innocuous change resulted in a product of slightly lesser purity that, alas, was significantly less stable. The latter fact came forward upon violent decomposition of $\sim$1 ton of product during the drying step. Subsequent investigation revealed that the process change product was somewhat less crystalline and had a significantly earlier onset of decomposition, such that at 50–55°C the self-heating process of decomposition started and rapidly took the material to its violent outcome.

Finally, thermochemical hazards assessment needs to start upon scale-up to the kilo lab, and if the structures at hand are suspect, some basic screening should be done even sooner. The effort then needs to continue as the process is developed and scaled up to the pilot plant, ensuring that significant process changes are not missed—a task of skillful vigilance, as the above example emphasizes.

2. Process Design from the Assessed Hazards and Achieving an Acceptable Risk

As indicated in Fig. 20, the hazards assessment data need to be placed in a process design context, in which scale issues arise forcibly: loss of surface to volume ratio, longer time cycles of certain batch events, more difficult mixing, larger in-process inventories, and many other. Upon scaled-up development, a reasonably specific design of the scaled-up operation needs to be challenged by the hazard and the resulting level of risk evaluated. This requires a sufficient engineering input and a deliberation commensurate with the magnitude of the hazard, and the exercise resembles the do-loops of computational code, with the effort resulting in a process design solution deemed to have an acceptable risk. Often enough this analysis leads to: (a) a significant change in the basic process (the scaled-up risk demands a lesser hazard) or (b) to a highly engineered design (the hazard is accepted, but its scaled-up risk is also accepted). Examples of these outcomes are the switch to a different reaction to get to the same structure or the use of a continuous reactor (or skipping the isolation of a dry unstable intermediate), respectively.

Beyond the above, more detailed methods of analysis exist for final plant design (e.g., HAZOPS or similar methods (30, pp. 42–178), with the objective of ferreting out the risks arising from the basic process hazards as well as all other overlapping hazards in the specific context for the process operation. In many cases, the specific risk analysis results in changes to the safe processing envelope so as to deal effectively with the risks. This result is depicted in Fig. 21, where a contoured envelope is adopted so as to place greater "distance" between the permissible ranges of process variables and the risks. For example, the mixed acid concentration in a hazardous nitration may be lowered to reduce significantly the risk of catastrophic failure by corrosion of the preferred (and existing) reactor vessel in the plant.

Eventually, this continuing exercise embraces all the issues of the context of choice: the intended production capacity, the intended operating space, the plant site location and their myriad specifics. This is one of the principal reasons for having a close collaboration between the bulk drug

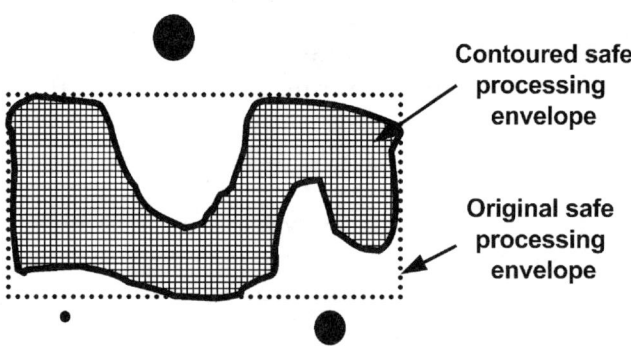

Figure 21 The contoured safe processing envelope. Detailed analysis of hazards in a specific context (specific process and plant designs) usually reveals specific risks, and implementation of the process in that context may require that the original safe processing envelope be contoured (i.e., modified for the said risks). In the figure, the original envelope has been shaped so as to create a distance from the risks that bears proportion to the magnitude of the risk, which is depicted by the size of each circular symbol for the risk.

development function and the process design function. The latter will usually be the most direct conduit to other functions at play, such as production planning, operational safety and environmental compliance. This desirable collaboration, however, does not imply that the bulk drug development team is a lesser participant in the process design effort; far from it, in the optimal scenario, the bulk drug development function possesses (and nurtures) sufficient process design skills.

3. Thermochemical Process Safety in Technology Transfer

By the time that formal technology transfer takes place, the thermochemical process safety issues are largely settled in the first manufacturing context, as described in the preceding paragraph. All that remains is the confirmation of the process performance and a reassessment of the risk on the basis of actual operating experience in the commercial plant. Of particular interest is how well the thermochemical process safety measures have been dovetailed with the full set of operational and environmental safety needs at the plant site, as this latter set, while preoccupied with a very broad range of hazards requiring very detailed measures (e.g., the lighting of exit signs, painted yellow strips, explosion proofing of electrical equipment, inertion of vessels with flammable materials, etc.), overlaps with thermochemical process safety (and depends on the hazards assessment data) on issues such as relief venting for runaway reactions, the lesser risk location of hazardous processing, emergency planning for chemical releases, etc.

B. Industrial Hygiene

Examples of serious harm to workers from materials used and made in the manufacturing workplace are well known: black lung disease and asbestosis stand out by the number of people affected and the severity of the results to long-term exposure to coal dust and asbestos fibers, respectively. Thus it is logical that, when dealing with bulk drugs (chemicals with potent and multiple biological activities), and as with thermochemical process safety, industrial hygiene (IH) issues arise early in the bulk drug process development cycle.

Unlike thermochemical process safety, however, the bulk of the IH effort (its hazards assessment and risk analysis) falls elsewhere, as the toxicology and preventive measure skills are not in the bulk drug process development function at all. Nevertheless, the bulk drug development function does have some important roles to play in ensuring the IH safety of its processes.

One of these roles is to be very alert to what chemicals are used (and their specific physicochemical properties) and ensure that the requisite toxicology screening is done on a timely basis, as well as tracking relevant process changes during development for their appropriate IH assessment. If anything, the bulk drug development team is very well positioned to be sensitive to the IH issues of new drug candidates given their knowledge of their biological activities and, at the very least, of their in vitro potencies. The interaction with the toxicology function also provides an early exposure to the toxicological profile of the bulk drug as it develops. Thus, if a compound's intended use is based on its cytotoxicity, that sets the stage for its handling even at the small bench scale. Indeed, there are toxicology screens that apply to the IH measures, since the latter are concerned not only with the bulk drug proper but also with all materials handled during the preparation or manufacture of the bulk drug. In many cases, intermediate compounds are found to have undesirable markers in these screens; e.g., dermal or ocular irritants, assorted acute toxicities, potential teratogenicity or mutagenicity, etc. The inevitable focus on the toxicology of the bulk drug must be balanced by ensuring the proper examination of the intermediates and to do so at an early stage. It is not possible, as an operating premise, to treat all compounds as worst cases; the appropriate data are needed for the protection of the personnel during development and for setting up the proper engineering measures and procedures for first manufacture.

The other role of the bulk drug team is, of course, to engineer major IH challenges out of the process, not unlike the elimination of thermochemical process hazards. For example, the process development effort may seek the avoidance of isolation of specific intermediates or of their handling as dry solids.

One might seek their fine particle distributions enlarged and, with the bulk drug proper, crystallization technology may be used to avoid milling of solids. Additionally, technologies of containment have been developed to deal with compounds of high potency in their biological activities at any scale of processing, and such technologies need to be practiced in the well-rounded bulk drug pilot plant, as indicated in Table 3.

Finally, access to sufficient IH skills is needed by the bulk drug process development function. Many IH issues are not in the scope of the toxicology function of drug development and require an additional set of skills, overlapping with operational safety and occupational health regulations as well. In one instance, a seemingly adequate containment of fumes and ventilation in an area where phosphorus pentachloride was handled could not prevent a very mild baseline irritation of respiratory mucosa of workers such that, upon their subsequent handling of a penicillin derivative in the next production campaign, severe allergic reactions developed. Nevertheless, the practitioner of bulk drug process development should be more than just aware of the IH issues and is hereby referred to some suitable introductory material (24, pp. 22–81).

C. Environmental Safety

Here we return to a forceful and decisive role of the bulk drug process development team. Just as the chemistry sets the scope of the thermochemical process hazards, so it does in setting the environmental profile of a bulk drug process: its inherent benevolence ("green" chemistry) in one extreme and its highly engineered implementation, made possible by intensive and extensive abatement and waste treatment measures as the other extreme.

Early assessment of environmental profile (or potential impact) is the best tool to steer the chemistry along a greener path. While the actual data are much harder to obtain for some key components of the profile (e.g., aquatic toxicity), suitable screens exist (31, pp. 93–177; 32). These, coupled with preliminary process design as to waste loads and some assumptions as to manufacturing sites, make it possible to feed back to the synthesis conception any of the following:

- The profile is such so as to merit the immediate search for alternatives for some specific aspects of the synthesis. The synthesis team may find such urging very disagreeable, but organic chemists have come a long way in accepting such judgments, even as early estimates. Of course, much depends on the level of skill and recognition of the assessors, which is one of the reasons for the early environmental assessment effort to be carried out by qualified people within the process development function, where they are generally perceived as less bureaucratic and regulation driven than comparably qualified people in a corporate or manufacturing function.
- The profile is promising and some particular aspects need adjustment or early environmental engineering attention.

Following the early assessment, the parallel with the thermochemical process safety effort is quite close, except for the greater difficulty and longer time cycle of some of the key data gathering. The issue of risk as a function of context, site location in particular, arises more sharply than with thermochemical risks due to the greater variety of downstream impact issues and of how far downstream they might arise. Questions of impact of the eventual discharges on seemingly remote and valuable habitats can arise, particularly with residual concentrations of highly potent drugs (e.g., mutagens, endocrine modifiers). More recently, the issues of drugs in drinking water sources and the ultimate fate of drugs excreted by patients have entered the regulatory expectations.

The companion Chapter 3 will revisit the environmental safety topic in the process design and technology transfer context. However, it can be stated herein that the environmental profile of the bulk drug process has moved up in the priorities of R&D as the drug dossier needs to address various levels of environmental safety assurances in individual regulatory submissions. The new drug application (NDA) in the United States, for example, requires an environmental impact statement of a scope that cannot be dismissed. It no longer suffices to provide statements of assurance as to compliance with all applicable environmental regulations.

Finally, there often are overlapping jurisdictions bearing on the ability of getting first manufacture started on a timely basis. All need to be satisfied that the intended manufacturing will not adversely affect the respective environments, just as communities and environmental advocacy organizations may need to be reassured. For all these and the above reasons, the environmental profile of the bulk drug process has risen in its importance, making it a good business choice to have a competent and well-quipped environmental technology function within the bulk drug development function and a close collaboration with the complementary environmental skills in the manufacturing and corporate organizations.

VII. OUTSOURCING IN BULK DRUG PROCESS DEVELOPMENT

The last decade has seen a drastic transformation of the bulk drug manufacturing milieu, including the adoption by the research-based drug industry of a business model (perhaps approached as a gospel in some cases) that greatly reduces the role of bulk drug manufacturing in-house and increasingly places it with outside suppliers. The latter have proliferated in the rush to capture a more profitable business that fine and specialty chemicals, and many are fully engaged in drug intermediates and bulk drug manufacturing, to the point that by 2002–2004 overcapacity exists.

Inevitably, this has had a major impact on the bulk drug process development as well. Given the manufacturing driver for this shift, the overall outsourcing topic will be discussed in the next chapter, including the said impact on the bulk process development.

VIII. IN CLOSING

Clearly, this chapter has described bulk drug process development as a complex, richly textured activity that is deeply rooted in scientific and engineering skill. The discussion has been largely based in the context of a large drug company where all the requisite skills reside, mostly in R&D, but complemented well by those of the downstream organizations.

The reader may well ask, particularly as the new drug business faces increased pressures to do it faster and in a more regulated environment, does the smaller organization have a chance to succeed? What if the seemingly indispensable critical mass is not there and, instead, the task must be done by dovetailing as best one can resources and functions from multiple organizations? To the author the answer is clear. Bulk drug process development is a business where size matters and matters greatly, and if success is measured by the timely introduction of new drugs (not just one drug at a time) on a broad marketing base, then the smaller organizations labor at a disadvantage and the virtual company struggles with projects of any scope. Indeed, for bulk drug projects of unusual technical difficulty, the smaller organization seems faced with insuperable odds. Yet, none of this denies the opportunity for the bulk drug process developer practitioner to excel and find professional fulfillment in any environment without regard to size; all that is needed is the requisite skill and dedication to one's work, as well as reaching out for the best possible collaborations that might be available.

REFERENCES

1. Pisano GP, Wheelwright SC. The new logic of high-tech. Harvard Bus Rev 1995; 73:93–105.

2. Gadamasetti KG. Process chemistry in the pharmaceutical industry: an overview. In: Gadamasetti KG, ed. Process Chemistry in the Pharmaceutical Industry. New York: Marcel Dekker, 1999:3–17.

3. Repič O. Principles of Process Research and Chemical Development in the Pharmaceutical Industry. New York: John Wiley & Sons, 1998.

4. Anderson NG. Practical Process Research and Development. San Diego, CA: Academic Press, 2000.

5. Lee S, Robinson G. Process Development—Fine Chemicals from Grams to Kilograms. Oxford, UK: Oxford University Press, 1995.

6. Atherton JH, Carpenter KJ. Process Development—Physicochemical Concepts. Oxford, UK: Oxford University Press, 1999.

7. Laird T, ed. Organic Process Research & Development. ISSN 1083–6160. Columbus, OH: American Chemical Society, 1997.

8. Flickinger MC, Drew SW. Encyclopedia of Bioprocess Technology. New York: John Wiley & Sons, 1999.

9. Calam CT. Process Development in Antibiotic Fermentations. Cambridge, UK: Cambridge University Press, 1987.

10. Atkinson B, Matuvina F. Biochemical Engineering and Biotechnology Handbook. 2nd ed. New York: Stockton Press, 1991.

11. Saunders J. Top Drugs—Top Synthetic Routes. Oxford, UK: Oxford University Press, 2000.

12. Corey EJ, Cheng X-M. The Logic of Chemical Synthesis. New York: John Wiley and Sons, 1998.

13. Shinkai I, et al. Tetrahedron Lett 1982; 23:4903–4910.

14. Lin JH, Ostovic D, Vacca JP. The story of Crixivan®, an HIV protease inhibitor. In: Borchardt RT, et al, eds. In: Integration of Pharmaceutical Discovery and Development—Case Histories. New York: Plenum Press, 1998:233–255.

15. Trost BM. The atom economy: a search for synthetic efficiency. Science 1991; 254:1471–1473.

16. Soderberg AC. Fermentation design. In: Vogel HC, ed. Fermentation and Biochemical Engineering Handbook. Park Ridge, NJ: Noyes Publications, 1983.

17. Verrall M, ed. Downstream Processing of Natural Products. A Practical Handbook. New York: Wiley, 1996.

18. Mathieu M. New Drug Development: A Regulatory View. 5th ed. Waltham, MA: Parexel, 2000.

19. Berry IR, Harpaz D. Validation of Bulk Pharmaceutical Chemicals. Buffalo Grove, IL: Interpharm Press, 1997.

20. Paul EL. Design of reaction systems for specialty organic chemicals. Chem Eng Sci 1998; 43:1773–1782.

21. Proctor LD, Wart AJ. Development of a continuous process for the industrial generation of diazomethane. Org Process R&D 2002; 6:884–892.

22. Martinelli MJ, Varie DL. Design and development of practical synthesis of LY228729. In: Gadamasetti KG, ed. Process Chemistry in the Pharmaceutical Industry. New York: Marcel Dekker, 1999:153–172.

23. McConville FX. The Real Pilot Plant Book. Worcester, MA: McConville, 2002.

24. Crowl DA, Louvar JF. Chemical Process Safety. Fundamentals with Applications. Englewood Cliffs, NJ: PTR Prentice-Hall, 1990:17–19.

25. Ramondetta M, Repossi A, eds. Seveso, 20 Years After. Milano, Italy: Fondazione Lombardia per l'Ambiente, 1998.

26. Stull DR. Fundamentals of fire and explosion. Am Inst Chem Eng (AIChE) Monograph Ser 1977; 73:10.

27. Center for Chemical Process Safety. Guidelines for Reactivity Evaluation and Application to Process Design. New York: AIChE, 1995:9–173.

28. Barton J, Rogers R. Chemical Reaction Hazards—a Guide to Safety. 2nd ed. Rugby, UK: Institution of Chemical Engineers, 1997:1–84.

29. Rowe SM. Thermal stability: a review of methods and interpretation of data. Org Process R&D 2002; 6:877–883.

30. Skelton B. Process Safety Analysis. An Introduction. Houston, TX: Gulf Publishing Company, 1997.

31. Allen DT, Shonnard DA. Green Engineering—Environmentally Conscious Design of Chemical Processes. Upper Saddle River, NJ: Prentice-Hall, 2002.

32. Venkataramani E, et al. Design of an expert system for early environmental assessment of manufacturing processes. Proceedings of the 43rd Purdue Industrial Waste Conference. Boca Raton, FL: Lewis Publishers, 1989.

3

Bulk Drugs: Process Design, Technology Transfer, and First Manufacture

CARLOS B. ROSAS

Rutgers University, New Brunswick, New Jersey, U.S.A.

I. INTRODUCTION

This chapter complements its preceding companion chapter 2, which addressed the task of bulk drug process development. The tasks addressed herein overlap the development of the process, as process design does, or culminate the development task, as technology transfer and first manufacture do. As in the previous chapter, this one seeks to provide a sound perspective of the latter tasks to the uninitiated and the new practitioner, while the structured presentation and the deliberately inserted points of view may interest and possibly challenge the experienced practitioner.

First, there is the promotion of deliberately overlapping the experimental development of the process with its design into a manufacturing plant. Valuable as it is, however, this overlap is often not used as a powerful method in seeking the better process and a manufacturing plant to match, but is practiced ineffectively, strictly as a necessity of the time-to-market imperative. Sometimes the jurisdictional divide at the development/design boundary is too deep; or there is an interdisciplinary gap, with chemists on one side and engineers on the other; or the process design becomes earnest too late to influence the development. Indeed, many scale-up difficulties cannot be identified or quantified soon enough without a sufficient process design effort that runs parallel and close to the development.

Then, there is the lessened character that the process design subdiscipline has developed as the result of many bulk drug projects being handled by design and construction firms, where the practice of process design can be unduly conservative, or too pliant to the client's wishes, or so lacking in the bulk drug processing skills so as to offer nothing beyond what the client brings to the project, with the client's errors or limitations dutifully incorporated into the design. In other projects, such as those that outsource manufacturing, the emphasis on process retrofit into existing plant is heavy and the process design, if any, is often beyond the grasp of the client. This harsh assessment is warranted by the penalties often paid, unknowingly at the time, because of the lack of the appropriate process design skills and practices in scaling up bulk drug processes, or simply by the absence of a mechanism to exploit the opportunities in deliberately overlapping process development and design.

Alas, chemical process design skills are hardly ever taught formally; the first and last academic exposure most engineering students have to the subject is a rather superficial and highly structured "process design" project at the undergraduate level. To make matters worse, computer software tools that can aid process design have usurped that undergraduate task, often reducing the student's effort to little more than filling blanks in fairly rigid templates, some times with proposed operational designs that can be hilarious (e.g., a *stirred* tank for a Kolbe reaction loaded with 4000 kg of 2-in steel balls!) and usually missing the learning experience of manipulating design options at the conceptual level. Yet, sound process design is a requisite of good process performance in the manufacturing plant, and creative process design is practically indispensable in achieving superior processes and plants, as well as in exploiting advantageous chemistry that might be difficult to implement in the plant. Thus the wisdom of fostering the formal development of those skills and the development/design overlap in industrial practice; placing emphasis on the conceptual and unstructured aspects, as these are not addressed well by the current computational aids that are widely used, and are less likely to be pursued aggressively by engineering design contractors.

Another objective of the chapter is to establish the value of another overlap: process design is, for all practical purposes, the first stage of technology transfer. Moving a bulk drug process from the development environment to that of first manufacture is a delicate task that is made more difficult without a competent process design component. Once these arguments are presented, the chapter is meant to flow rather naturally as a series of annotated common sense prescriptions for sound technology transfer, proven to the author over numerous projects and observations that ranged wide through the practice of bulk drug chemical processing. We will dwell on these measures as they apply through first manufacture, setting the stage for mature manufacture as a function of product growth.

The chapter also includes, in closing, some observations on the technologies of bulk drug processing—development as well as manufacture. While perhaps couched as pearls of wisdom, they reflect some of the deeply held views of the author that could help the new practitioner with a perspective of the bulk drug business enterprise.

II. THE PROCESS DESIGN TASK IN BULK DRUGS

A. Definition and Scope of Process Design

Process design is, first of all, *not* the development of the precise specifications for a performing chemical plant, whether built on a green field or merely the modification of an existing facility to accept a new process. Instead, process design takes place well before such specifications can be drawn and it is only through its completion that the *plant design* (in contrast to the process design) can be carried out, defining the future plant to the extent that equipment can be procured and installed or an existing plant modified. Necessarily, process design has a much broader scope, including a largely conceptual component that comes about early in the overall effort, confronting issues and unknowns in a sequence that is outlined in the following sets of questions:

1. Broad brush definition of the task and its probable capital cost, venues and timetable.

- At what approximate output will the process be first run and when? On what operating basis?
- What manufacturing cost can we project? What are the top cost reduction targets?
- What are the probable materials and energy balances of the process?
- Which operating site makes the most sense? In what operating area of the site of choice?
- Should it be run in a new plant or a retrofit?
- What are the prospects for outsourcing some or all of the manufacturing tasks? Which tasks are most likely to be outsourced? When will that decision be made?
- What is the probable range of capital cost? Is the high end of the range acceptable? Is the low end (via outsourcing) attractive enough?
- What is the probable timetable to an operational plant? Is the timetable acceptable?

From dealing with these and other broad questions early, the process design should continue without pause as the features of the process take shape with the benefit of feed-back, and the issues of site selection and outsourcing become more distinct. Indeed, as Fig. 1 illustrates, the fundamental objective of the parallel exercise with development is for the latter to *reflect the teachings of the process design.*

2. Definition of the various process steps as they might be operated in the most probable venue of choice. For example:

- Will the process run as currently operated at the pilot scale? At what scale? Batch-wise? Continuously?
- Integrated for optimal layout or placed opportunistically throughout the existing plant?

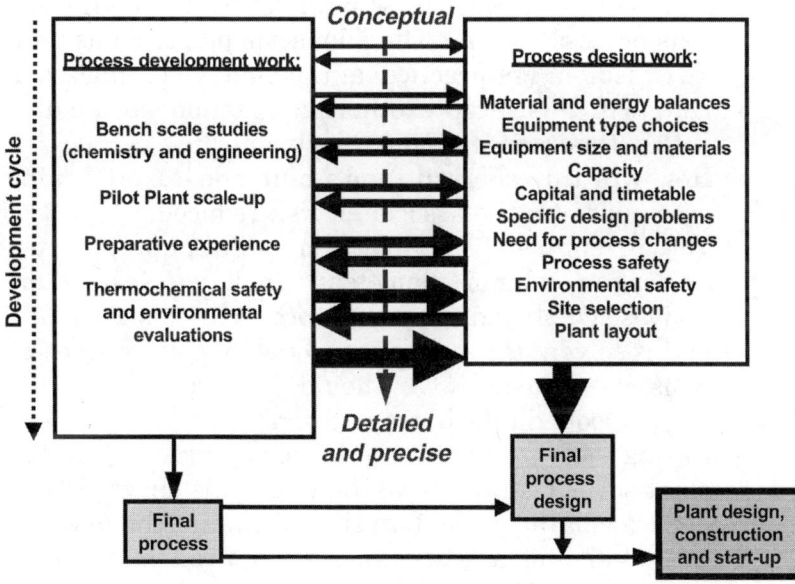

Figure 1 Flow chart of the process design effort—bulk drugs: The overlap with the process development effort provides an unmatched opportunity to seek the better process by using the feedback that process design can provide. Both efforts move from the conceptual to the detailed and precise that is eventually needed to permit plant design, construction, and start-up. Similarly, as indicated by the horizontal arrows between the two efforts, both feed forward and feedback improve in defining the evolving process and its design as the efforts take place.

- With or without solvent recovery or recycle?
- Are the environmental burdens acceptable?
- Are there waste or pollution issues demanding at-source treatment?
- What are the identifiable risks from the know process hazards? Where are the safe processing boundaries (i.e., the safe processing envelope described in Fig. 19 of Chapter 2)?
- First flow sheets and their material and energy balances are defined. What adjustments do they suggest? To the design or to the process?

3. More specific design issues arise. For example:

- How will these solids be separated from the highly viscous process stream? The pilot scale practice has been an expedient not practical at the manufacturing scale.
- How will this large exothermic reaction be handled within the residence time constraints?
- How will this aberrant and unintended exotherm be precluded? Or the associated risk reduced?
- The current method of isolation requires unprecedented adsorption column diameters. Shall we seek alternatives? Or should carry out more scale-up studies?
- Is this solvent throughput reasonable? Can concentrations be adjusted? Or should we use an internal recycle loop via flash evaporation?
- Drying these particular solids seems intractable at the plant scale. Can we move the wet solids forward? Use a more volatile solvent wash to facilitate the drying?
- Availability of a vessel in the required material of construction is a problem. Do we seek alternative materials or alternative processing conditions?
- The industrial hygiene data for this intermediate demands a given level of containment. Can the isolation be avoided?

Clearly, a myriad of such questions arise, preferably sooner rather than later. There will not be satisfactory answers to some, which should trigger subsequent iterations in the feedback loop shown in Fig. 1. Although some extent of feedback from manufacturing planning always occurs, it is best by far to establish the feedback loop as early as possible and in a framework that makes the process development effort sensitive to it. In other words, there must be responsiveness within the development effort, as indicated in Fig. 1; particularly when the issues that arise are inconvenient or undermine the more basic process choices that might have been made or that have the greatest appeal to the development team.

The best conditions exist where there is a process design capability that collaborates with the process development team; indeed, the latter should participate by virtue of having a modicum of skills to understand the process design feedback, as well as having the wisdom to act as needed. Conversely, the process design team must have sufficient knowledge of chemical processing at large to understand the process imperatives or the rationale that makes a given process approach so attractive so as to accept the design challenges. Frequent communication, even if at times burdensome, and overlapping efforts are the key components of successful process design for a developing process. A new process, even if presented as developed, must survive the challenge of its process design; thus the compelling rationale for the overlap of the two tasks, as shown in Fig. 1. Another depiction of the results of the overlap of process development and design is given in Fig. 2 for the example of a multistep process under development, showing the range of what happens as process design takes place: from straightforward implementation (Step 1) to iterative effort requiring change of the process concepts (Step 4) as indicated by the use of different letters and superscripts.

After considerable evolution, the principal finished product of process design is the process and instrumentation diagram (P&ID), eventually issued in what is usually called the approved for construction version (AFC), the definitive successor of various intermediate versions and their revisions. A slice of such a P&ID is shown, with some simplifications, in Fig. 3. Obviously, there is a great deal of supporting detail

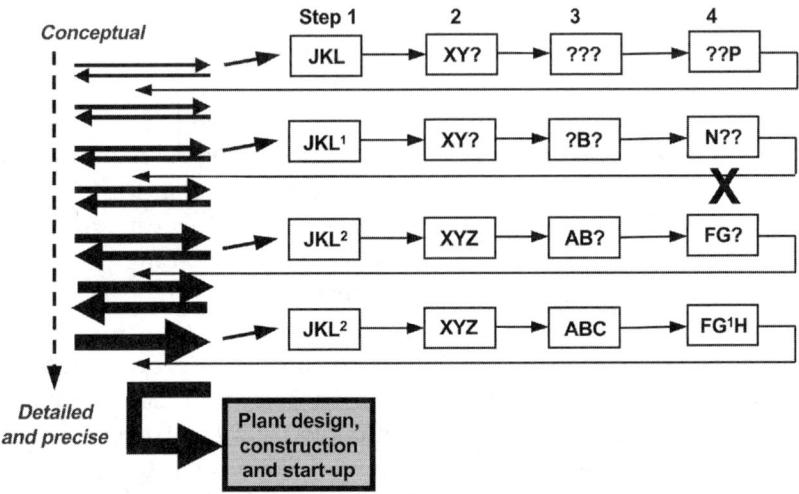

Figure 2 Process design of a developing bulk drug process: Along the same lines of Fig. 1, process approaches (chemistry included), are indicated by the first letter in each box, versions within the approach are indicated by the subsequent letters and elaborations of a version by the superscripts, all used to depict the optimal evolution of a process through the continuous feed forward and feedback between development and process design. Note that Step 1 moved forward with little change, whereas in Step 4 a completely different process approach was found necessary.

that attaches to the P&ID AFC, but a critical examination of the diagram is the core of the subsequent plant design effort.

Perhaps the most difficult aspect of overlapping the development and design for a chemical process in flux is that of the uncertainties from unsettled process issues, which may range, depending on the character of the process development organization, up to the actual chemistry for part of the process. These uncertainties often appear in the P&ID as blank or ill-defined areas under the heading of "to be determined", as depicted in Fig. 4. Such uncertainties arise in the projects with the most technical difficulty or in the fortunate instances when a clearly superior and compelling part of the process exists but comes late to the fore, thus presenting the most severe test to the skill and discipline of the development/

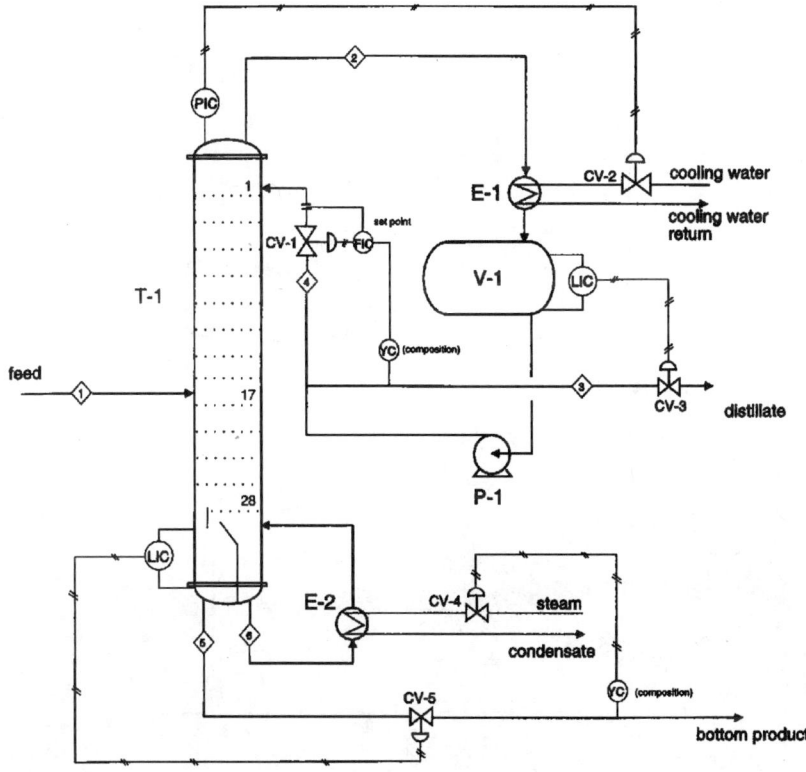

Figure 3 The process and instrumentation diagram (P&ID): Shown as a simplified version stripped of some detail (to permit an uncluttered and legible diagram for reproduction herein), the P&ID identifies each item of processing equipment, their connectivity and the control instrumentation loops. Not shown, but generally present in the final approved for construction version of the P&ID are the details of the piping, materials of construction, pump capacities, etc. Obviously, there is a wealth of other material that accompanies the P&ID, but the latter is the centerpiece of the process design package used to execute the plant design.

design interaction. Organizations or interacting teams that can manage those situations under the time-to-market compulsion have a major tactical advantage and, if they can exploit them under an enlightened R&D management, their advantage can be strategic.

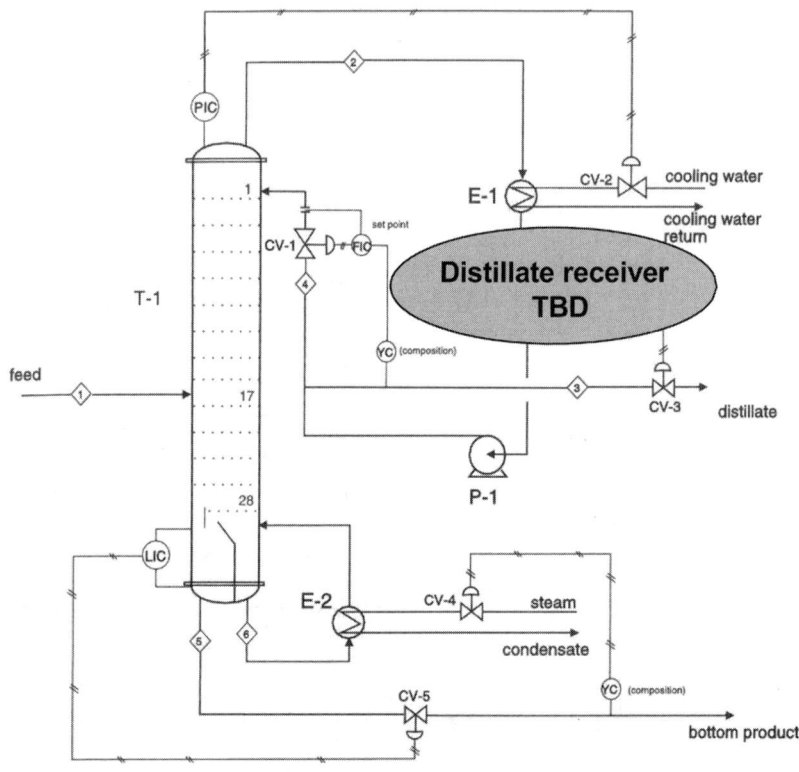

Figure 4 The "to be determined" (TBD) provision in a P&ID: Although a very modest example is shown, the TBD provision is used as needed to indicate parts of the process design that may trail the overall design. This provision is particularly useful when the process design needs to move rapidly, sometimes at some risk that the outcome of the TBD item may require redoing some of the related design work.

There is, of course, the plant design effort, requiring a level of detail that far exceeds that of process design and that follows it with considerable overlap, as depicted in Fig. 5. We should note herein that a bad process design cannot be turned into a good one by plant design means, as the latter are aimed at faithful implementation of the process design intent.

Finally, it should be understood that the discipline explicit in the above prescriptions does not depend all that much on the size of the organization; it can be applied by the

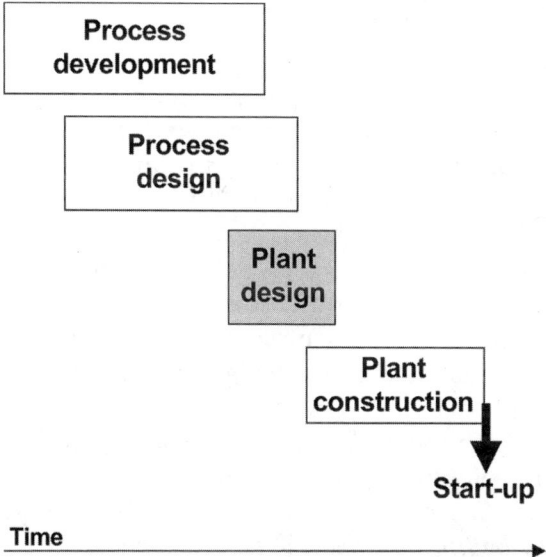

Figure 5   Plant design as the sequel to process design: The ability to compress the overall time cycle to first manufacture also depends on the extent to which process and plant design can overlap and permit earliest plant construction activity. The instance of retrofit into existing plant, although generally providing for the shortest cycle, can be adversely affected if the development and design overlaps miss an essential equipment item that is not at hand.

smallest team as long as the requisite complement of skills exists within it or is sufficiently accessible and responsive elsewhere. Indeed, one of the various arguments for having a core of engineering skills within bulk drug chemical development is the ability to take snapshots of the developing process and do a good deal of informal and intimate process design or *"back-of-an-envelope"* design—e.g., is this the manufacturing plant we wish to operate? Or would this plant be amenable to ready expansion? Such snapshots permit swift sifting of approaches and choices, more rapidly adjust the bench and pilot efforts, and spare formal process design effort for more mature versions of the process. For projects with capital or product cost sensitivities, the snapshots also permit rapid estimation of the alternatives.

B. Process Design as the First Stage of Technology Transfer

Examination of Figs. 1 and 2 confirms the previous assertion that process design is the first stage of technology transfer or, at the very least, provides the opportunity to initiate technology transfer to the advantage of the project. This is because in most drug manufacturers the process design function (to whatever extent it applies to the project—new plant or retrofit) is associated with the first manufacture of the product, and the know-how of the developing process begins to reach the operational organization as process design begins. In both figures, the increasingly heavier arrows, as the project progresses, indicates the know-how flow and the corresponding feedback across the development/design boundary. This fact is also apparent in Fig. 6, where process design begins when the process know-how *begins to take its eventual shape.*

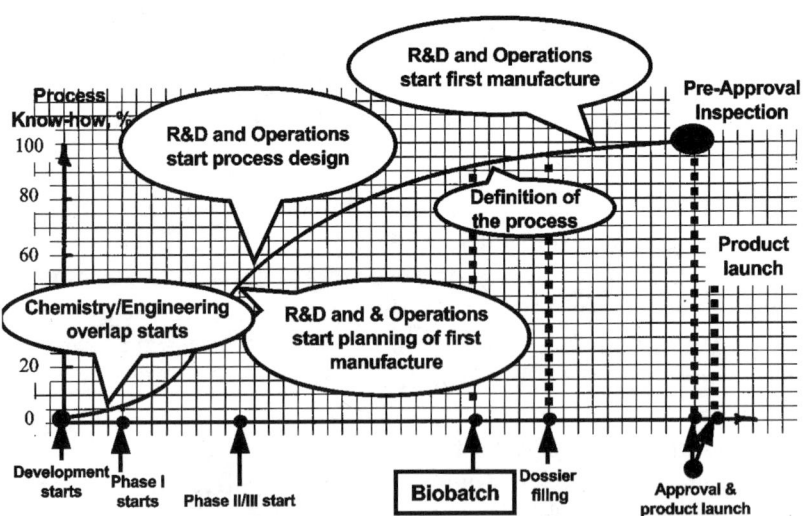

Figure 6   An eagle's eye view—from process development to plant start-up: The earliest start of planning and process design for manufacture as overlapping activities with process development provides the best chance of earliest first manufacture.

Participation in the development learning curve, reasonably close for the process design function and somewhat distant for the manufacturing function, can be very beneficial to the overall project, albeit at times the inevitable vicissitudes of process development cause distracting anxieties on the downstream side of the development team—an occasion for the appropriate managements to becalm the situation. Nevertheless, the opportunity for the operational side to prepare for the technology transfer and first manufacture is excellent, and an exemplary mechanism for such transfer will be presented in Section III below.

However, the current prevalence of outsourcing and the frequent use of engineering design firms has created an environment in which technology transfer takes place in a variety of ways and some times not at all, as we will discuss. Alas, the tidy arrangement of doing everything in-house, as illustrated in the said figures, is gradually giving way to drug companies that manufacture only the very last stages of the chemical process. Nevertheless, the principles also illustrated in those figures are sound indeed, and good efforts to incorporate as much of them into whatever development/design or development/first manufacture boundary applies are worthwhile.

C. The Process Design Demands on the Process Body of Knowledge

The demands of overlapping process design with the development of the chemical process are more immediate and somewhat less rigorous than those of the dossier and of the final process design or the formal technology transfer events. Instead, good communications are essential (again those back and forth arrows in Figs. 1 and 2), preferably complemented with brief written reports when necessary. Apart from clarity and accuracy, timeliness is the next most precious quality of the information exchange that undergirds the overlap of process development and design. In other words, the snapshots of the developing process need to be rapidly examined through process design as required by the scope of the changes being introduced or contemplated.

Beyond those demands, the process design function will eventually need the complete and fully organized body of knowledge (Table 4 of Chapter 2) so as to permit:

a. process design;
b. plant design (project engineering design and construction);
c. procurement of materials;
d. preparation of start-up plans and operating procedures;
e. assessment of the process safety issues in the specific context of the plant: operational safety, industrial hygiene, thermochemical, and environmental safety;
f. assembly (and timely approval) of environmental and other regulatory permits;
g. definition of the process start-up targets of yield, capacity, waste loads, etc.;
h. dealing with assorted other matters, such as those arising from the plant's insurance, etc.

Aside from the procurement of laboratory equipment through the capital project, the information needed for the transfer of the in-process and QC analytical methods need not go through process design and may pass directly to the manufacturing organization (QA/QC included).

Figure 7 depicts a proven model for the flow of the process body of knowledge as it is applied to process design and first manufacture.

III. TECHNOLOGY TRANSFER OF THE BULK DRUG PROCESS AND FIRST MANUFACTURE

A. Definition and Scope of the Technology Transfer

Technology transfer has become the term that more appropriately describes all the events associated with the first manufacture of a new bulk drug (or for that matter, of any new product with its own distinct process for manufacture,

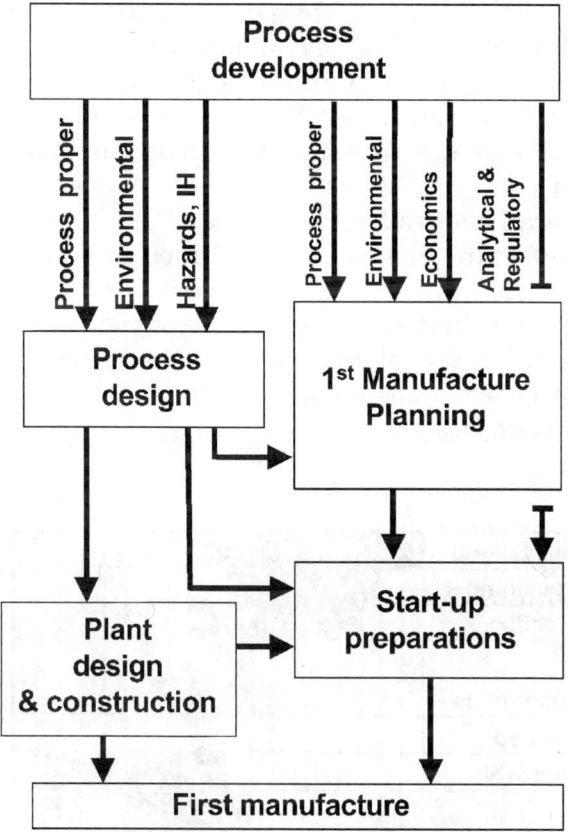

Figure 7 A model for the flow of the process body of knowledge: Successful first manufacture (timely, sufficient, and reliable) depends not only on the assembly of the requisite process body of knowledge, but also on its timely flow to the various downstream activities. The model shown above has worked well in each of numerous instances that the author has seen it applied.

which will usually come from without the manufacturing environment). Older and not-so-old practitioners are probably more comfortable with less comprehensive, but very descriptive terms, such as "process start-up" or "process demonstration."

In its broadest definition, so as to capture all the activities for its execution, the scope of bulk drug technology transfer encompasses the tasks listed below. (Here the reader is encouraged to place bookmarks on Figs. 11 and 13 of Chapter 2 for perusal, as well as review the approximate Gantt chart for the overall technology transfer in Fig. 8 below.)

Early stage (around the time the new drug candidate enters development):

(a) The operations area acknowledges the task of a probable first manufacture tied to an expectation of regulatory approvals to market a new drug.

(b) At the same time, the process development function acknowledges its share of the above task—providing the technology for the safe, dependable and timely execution of the first manufacture.

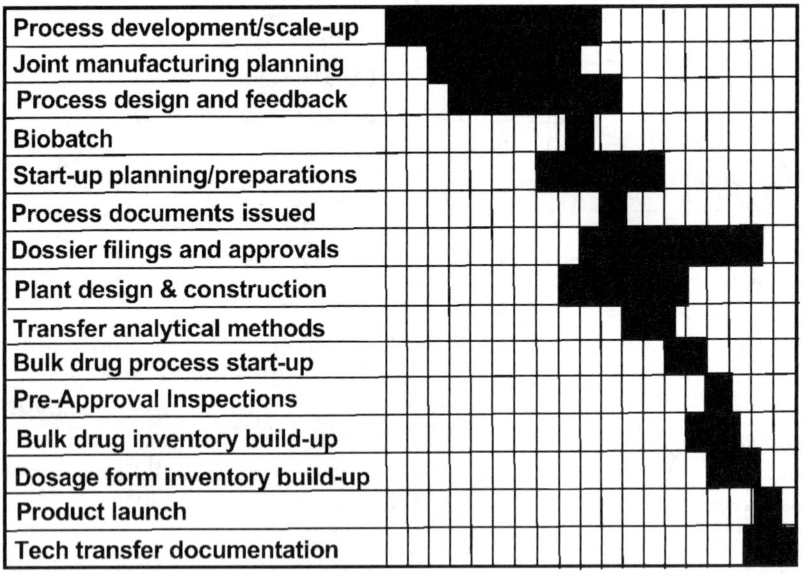

| Process development/scale-up |
| Joint manufacturing planning |
| Process design and feedback |
| Biobatch |
| Start-up planning/preparations |
| Process documents issued |
| Dossier filings and approvals |
| Plant design & construction |
| Transfer analytical methods |
| Bulk drug process start-up |
| Pre-Approval Inspections |
| Bulk drug inventory build-up |
| Dosage form inventory build-up |
| Product launch |
| Tech transfer documentation |

Figure 8 Gantt chart for the technology transfer of the chemical process for a new bulk drug: The chart shows all of the key components of the overall task, with an approximate indication of their relative positions on the time-to-market cycle and, to a less precise extent, of the relative widths of their timelines; the latter can vary considerably as a function of the process scope, new plant vs. retrofit into existing plant and in-house *vs.* extent of outsourcing.

(c) The need for probable capital approvals and subsequent expenditures is forecast, presumably within the parameters of an established longer-range plan that includes the launch of the new drug product as a probable event.

Next stage:

(d) The process development function begins its collaboration with the operations area in addressing the broad brush definitions of the project (as set out in II.A.1 above).

Next stage:

(e) Overlapping process development, design and manufacturing planning takes place.

(f) Capital approvals are sought, in portions and from a range forecasted for the total project.

(g) Starting material sources are developed and business terms negotiated (this may include extensive outsourcing of, say, intermediates manufacture).

Next stage (some time after the biobatch milestone, but before filings from the dossier take place):

(h) The process body of knowledge is documented.

(i) The final process design is completed.

(j) The plant design is completed and installation work proceeds.

(k) Starting materials and auxiliaries are purchased.

(l) Start-up plans and operating procedures are developed.

(m) In-process, QC, and regulatory methods (stability) are transferred.

(n) Process safety issues in the specific context of the plant are settled: operational safety, industrial hygiene, thermochemical, and environmental safety.

(o) Environmental and other regulatory permits are assembled, filed, and approvals obtained.

(p) Definition of the process start-up targets of yield, product quality, capacity, waste loads, etc.

(q) The process validation plan is defined.

(r) The process start-up team is assembled.

(s) The plant installation is tested and readied for the process.

(t) The process is started up and validated.

(u) Preapproval inspections take place.

(v) Process consolidation—the start-up continues to demonstrate all targets under (p).

(w) Results are documented, including updates operating procedures, in-process controls, etc. Heads of the start-up team sign off.

(x) Mechanical/instrumentation items punch-list and the "To Do" list are prepared.

(y) The start-up team is disbanded, but liaison persons are designated for matters arising. Manufacturing takes over.

The reader should beware that hidden within the above reassuring list are all the necessary actions to solve unexpected problems, particularly those arising during process start-up, validation, and consolidation, or those in response to significant observations from preapproval inspections. Difficulties in technology transfer are inevitable; no such large number of activities that must dovetail precisely can go without some adversity or something being overlooked. Yet, *well-executed* projects for complex chemical processes generally meet their targets of bulk drug deliveries and the existing process body of knowledge and assembled resources permit the swift resolution of arising difficulties.

An approximate sequence of events *following* the designated validation work (and its follow-up) that is useful for planning purposes is (assumes a multistep process of significant scope):

> During validation—25% of design capacity is reached.
>
> Month 2—60% of design capacity is reached.
>
> Month 3—80% of design capacity is reached. All lots are without quality issues.
>
> Month 4—100% of design capacity is reached. All lots are without quality issues.
>
> Month 5—procedures updated. Summary memorandum on results is signed off and issued.
>
> Month 6—Operating personnel training confirmed (but meant to continue). Comprehensive process start-up document issued.

Obviously, the above timetable will vary with the scope of the process start-up: number of steps, number of plant sites, intrinsic process complexity, time cycle (e.g., long fermentation cycles plus downstream processing plus any semisynthesis to follow). Sensible allowances to the above figures should be made.

B. Mechanisms for Technology Transfer

There are as many technology transfer mechanisms as there are operational arrangements in bulk drug manufacturing. However, the practice of technology transfer within research-based drug companies can be said to take place within either of two environments:

1. R&D driven. In this arrangement, the R&D division delivers, through its process development organization, a complete process to the operations division, which generally uses its process design function as the *principal* gate to receive the process. While some specifics may vary from instance to instance, the technology transfer (as just defined under III.A) takes place along the following lines:

    R&D bears the principal responsibility for the technical success—it *demonstrates* its process to the operations division. Accordingly, R&D leads the effort and casts a heavier vote on decisions bearing on the process and its operation, not unlike a first among equals. R&D also acts as the technical liaison with contract manufacturers and transfers its process or chemistry to them, and eventually sponsors the suitability of materials from those contract manufacturers.

This environment offers the decisive advantage of *a single handover*—from the development activity in R&D to a performing plant that delivers bulk drug as required, and with reasonably well-defined responsibilities. Indeed, technology transfer is an activity with all the vulnerabilities of a handover, and the analogy with certain sports is quite apt, thus the advantage of a single transfer. Figure 9 attempts to describe the R&D-driven environment for technology transfer.

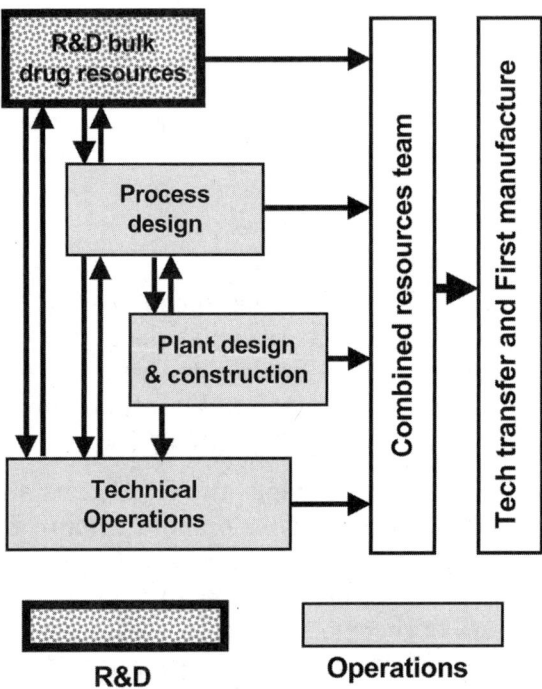

R&D   Operations

Figure 9 Technology transfer mechanisms—R&D driven: All of the process know-how flows from R&D organization to operations (manufacturing), outsources or both, although the diagram depicts the in-house case.

2. Stage-wise. In this arrangement, the process development is split along disciplines or along operational lines, causing more than one handover and with a greater spread of the technical responsibilities (Fig. 10):

a. The synthesis (or biosynthesis) and its analytical components are delivered by the R&D division (chemists and microbiologists only) to an arm of the operating division that has engineering and pilot plant resources. Obviously, the delivery of the chemistry must be in stages as it develops, as the bulk drug supply to drug development may be the responsibility of the operations group. Eventually, the latter transfers the process to manufacturing.

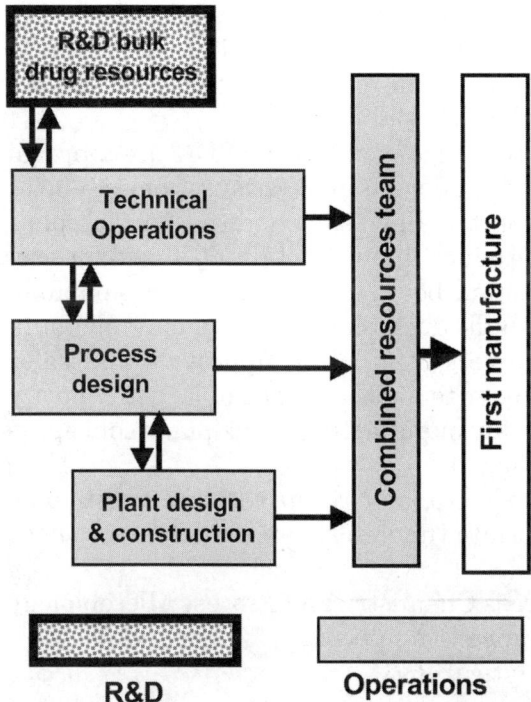

Figure 10 Technology transfer mechanisms—stage-wise across operational boundaries: Either a partially developed process or a developed process (through the last intermediate material) is transferred to the technical arm of the operations area, which carries out the first manufacture activity, with R&D playing a secondary or contingent role during the latter. Most often this mechanism results in two technology transfers.

b. The R&D division delivers the process to the operations area at some intermediate stages of development at which the chemistry has been established, with the rest of the development completed in the operations area. The developmental bulk drug supply is a shared responsibility, allocated according to ownership or control of pilot scale resources.

Regardless of the mechanism, however, all the activities under III. A need to be carried out, even if less tidily. The same is true for those frequent cases in which the sequence of process steps is divided among more than one manufacturing site. Even less tidily, the same is true for projects with significant outsourcing, certainly to the extent that the processes are reliably established at each supplier.

In recent years, the advent of the USFDA preapproval inspection method (and analogous inspections from agencies outside of the United States) have spawned the "launch platform plant"— a multiproduct chemical plant designed for versatility, faster turn-around between products and capable of manufacturing a new bulk drug (or more than one new bulk drug) until the full range of regulatory approvals and sales growth justify the transfer to another plant of larger capacity as a longer-term home for manufacture. This plant concept is discussed further in Fig. 11.

As to the technology transfer resources that must come together at the appropriate time, they range wide across both R&D and operations:

From R&D—Process Chemistry and Process Microbiology
    Chemical Engineering
    Analytical R&D
    CMC team that prepared the process input to
    the dossier

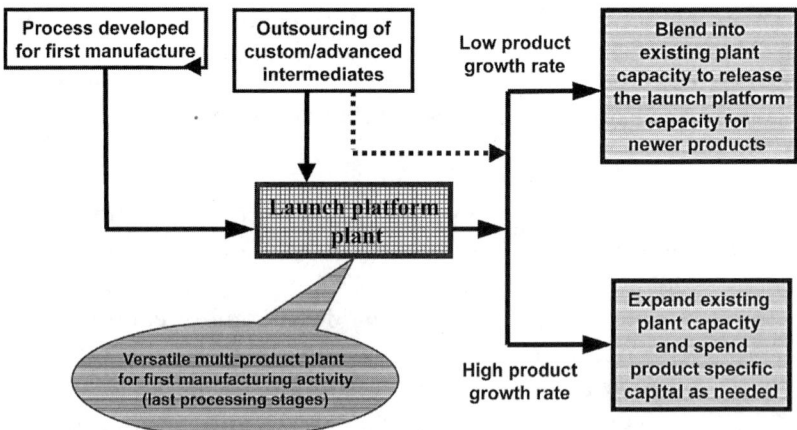

Figure 11 The launch platform plant: A device aimed at avoiding a significant plant construction task by retrofit of the new bulk drug process into an existing plant dedicated to first manufacture only.

From Operations—Process Design
> Plant design and installation (and
> their contractors)
> Technical Services (divisional or from
> the site of manufacturing)
> Production (from the manufacturing
> site)
> Plant Engineering
> QA/QC
> Operational, Health, and Environ-
> mental Safety
> Materials Management (divisional
> and site of manufacturing)

While some of the above players carry out crucial support roles and are active in the day-to-day effort of starting-up a new process, the principal burden rests with those with their hands on the actual operation, the process designers and the immediate laboratory support (testing, ad hoc experiments to obtain a missing datum, validate a hypothesis on a problem, or to run a process manipulation in parallel to the plant). They have not only the task of demonstrating the process in new or modified equipment but also that of training the operating personnel.

Some common sense and well-proven prescriptions are:

- A detailed log of events, preferably in clear English prose and with comprehensive entries, is essential. The "manufacturing operating instructions" or whatever formal record of the processing is created will generally be far too structured as a series of instructions and blanks for data and signatures, ditto for logs from the control computer system, if any.
- As suggested above, the plant Technical Services labs should be dedicated to support the process start-up around the clock in speedy and unstructured ways that QC cannot.
- The process start-up team should be well staffed in numbers and in the representation of all the skills

and experience accumulated during the development
and process design, set to apply more than sufficient
power to the task and *make rescue missions unnecessary.*

- All background documentation, from the development, the process design and the installation should
  be at hand and well organized for swift location of
  needed information.

- Operations management should keep its oversight
  discreet and be disciplined with respect to distracting
  the start-up team for the latest, particularly at times
  of stress.

- There should be at least one review meeting a day,
  attended by the principals of development, process
  design, analytical, technical services, production and
  project engineering (installation), run sharply and to
  the point, with "who does what by when" unambiguously defined.

- Data and other trends should be followed, preferably
  from some premeditated plan, so that the direction
  of process performance can be assessed soonest.

Finally, a technology transfer device that works very
well in avoiding pitched battles upon process start-up
problems—where are the funds to fix the problems?—is to
have a fixed amount of the capital budget for the project allocated as a contingency to such fixes and *under the control of
the start-up team.* This protects the process start-up from a
shortfall of funds due to underestimation of the original
installed cost of the plant. A total of 5% of the total capital
budget is usually a sound allocation for such contingency.
The latter is, of course, apart from the usual 15% contingency
of such capital projects, and is given back to the corporation if
not used (by keeping it out of the grasp of the plant manager).

C. Technology Transfer in the Outsourcing and Licensing Environment

Up to this point, we have alluded to the outsourcing environment while describing systems and prescribing ways of

operating largely in the context of a big pharma organization, where all the capabilities exist under, ultimately, a single management. This has been useful in that it has permitted presenting the tasks of process development and technology transfer comprehensively. Indeed, the tasks at hand are basically the same and need to be done just as well without regard to how the tasks might be divided between the various parties in the current outsourcing environment for bulk drugs manufacture.

We need, however, to understand the complexities added to the basic tasks by their becoming divided among:

- customers, ranging from big pharma to virtual and "almost virtual" drug companies;
- suppliers, ranging from large fine chemical manufacturers to small new companies with a claim to some niche processing technology;
- service providers offering to take only the customer's compound structure and do it all through first manufacture (generally relatively new and small companies), thus appealing most directly to the virtual and almost virtual customers.

Virtual drug companies are those that generally possess nothing more than the patents or the licenses to a compound, and operate by contracting all subsequent tasks. Almost virtual companies are those that, although having discovery and some clinical development capabilities, lack everything else.

Starting with the big pharma customer, the technology transfer task takes place in a relatively narrow range defined by the following poles:

- The customer wishes to outsource part or all of the bulk drug manufacturing, using one or more contract manufacturers. The customer also brings sufficient process know-how and assumes the responsibility of demonstrating its performance in the manufacturer's plant, according to a *well-defined process start-up plan* jointly developed by the two parties (1). Figure 12 describes this happy set of circumstances, clearly the best scenario for

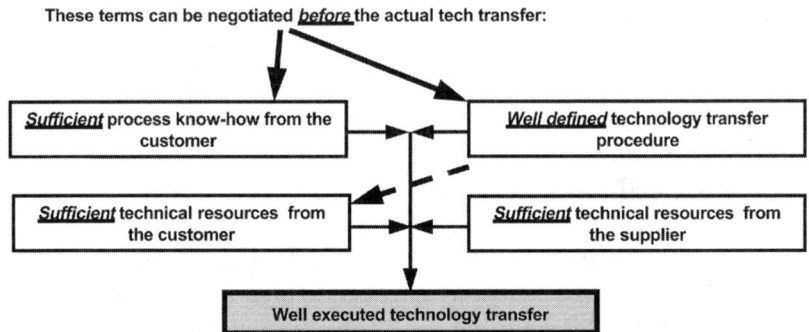

Figure 12 Technology transfer in the outsourcing environment—
the optimal scenario, shown on the basis of the well-managed transfer of sufficient process known to a capable recipient.

technology transfer between the customer (who has the
technology) and the contractor (who as the plant and the
right set of manufacturing conditions).

- The customer and the contractor come together on the
  presumption that the contractor's processing skills,
  plant capacity, and existing chemistry operations
  (with its available intermediates) constitute a significant advantage (time and cost) as a supplier of a given
  compound, generally an advanced intermediate. The
  customer may contribute all or part of the process
  for the conversion of the contractor's intermediate to
  the customer's target, although often enough the contractor does contribute the actual process.

Both poles define a range in which technology transfer is
greatly facilitated by the existence of a sufficiently developed process from the customer (in the first case, Fig. 12)
or, in the second case, by the relative ease in reaching
the target structure from the contractor's existing intermediate, with or without some process development collaboration.

However, on the other end of the spectrum, the virtual
and almost virtual customers have no process technology of
their own and thus seek one or more contractors that will

make the compound on the basis of their existing offerings of proximate intermediates. Here the customer has no technology to transfer and the matter resides entirely with the contractors. There is, however, a scenario that is increasingly found in the virtual and almost virtual customer domain and that can be best described as the "technology transfer from hell," which a mere examination of Fig. 13 will confirm.

Most cases of a compound being licensed to a virtual or almost virtual company bring no process technology (beyond medicinal chemistry and preliminary preparations) or partially developed processes. Thus, the licensor is, at best, in a weak position to transfer useful or complete process technology and usually lacks motivation beyond the precise letter of the license. Also in most such cases, the virtual or almost virtual customer lacks processing background or experience in the fine chemicals milieu, or hires people with such competency much too late.

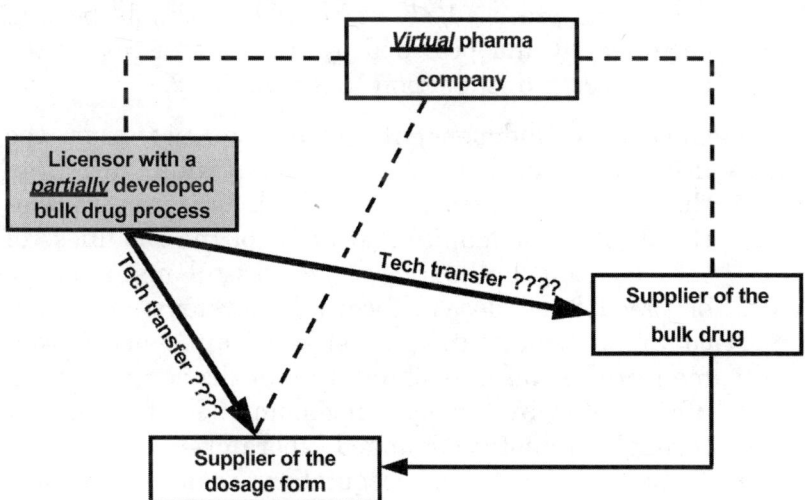

Figure 13 Technology transfer in the outsourcing environment— the worst scenario, shown on the basis of a virtual company that licenses a compound that does not bring a sufficient process body of knowledge.

While some of the above difficulties are inherent to the virtual character of the customer, the latter can take some actions to avoid the "technology transfer from hell." Namely:

- Hire or engage competent people in chemical manufacture, preferably with experience in dealing with fine chemicals manufacturers, and preferably right at the outset of mounting a serious clinical effort.
- Seek license language that unambiguously obligates the licensor to provide full documentation of the process and its experience with it (in a manner suitable for input to a dossier), as well as an iron-clad obligation to a serious technology transfer effort. Use milestone payments as a means to motivate the licensor on the latter.

  The licensee should beware of accepting the licensor's "production documents" as the core of the process documentation, as such documents—operating procedures with the blanks filled in—are poor vehicles for imparting process knowledge. To the extent that it exists, the process body of knowledge should be well documented and provided by the licensor, along the lines described in Section V of Chapter 2.

- Obtain an independent evaluation (not from the licensor or the potential contract manufacturers) of the development status of the bulk drug process relative to its first manufacture *and* of the readiness of the resulting bulk drug for successful manufacture of the desired dosage forms. If development is not complete, seek its completion in a competent environment rather than patching it up at the contract manufacturer or by hiring a modicum of staff to rush it through in hastily arranged laboratories.
- Manage its QA and regulatory team as *earnest participants* in establishing the bulk drug activity rather than approaching it strictly as a *policing task*.

D. Regulatory Aspects of Technology Transfer

The principal regulatory task in the technology transfer of a bulk drug process is its reduction to practice in accord with the process defined in the dossier, and to develop and execute a sufficient (not an excessive) validation plan. It is also important to do so at a contractor or contractors that understand the basis and procedures of the drug master file system and their obligations to the customer's dossier; having a GMP status is not sufficient, as the latter is the minimum requirement. The customer should scrutinize those aspects of the contractor's operations early in the due diligence process, including its change control procedures.

Prepared for preapproval inspections (PAI) is not a trivial matter, as significant observations may delay the approval of the corresponding applications; usually the PAI takes place after all other aspects of the application have been reviewed and found suitable for approval. Although GMP issues may arise, the principal objective of the PAI is to determine the soundness of the manufacturing process relative to the process described in the application. Some golden rules that can do no harm are:

- Be prepared to credibly answer questions on the spot (all knowledgeable personnel immediately available, all documents organized and handy). Questions should not linger unanswered because of not being prepared.
- Assume the inspector wants to know if you know what you are doing and that the scientific and technical background of the process has been mastered by those who will run it.

E. Transition to Mature Manufacture

Successful transition to mature manufacture (different from static or declining manufacture!) requires, of course, the firm basis of a sound and well-documented technology transfer, including a list of all the "To Do" items (those actions of modest scope that would consolidate and improve the reliability of the process and its operation). The process training of the operating personnel also needs to be consolidated, avoiding the feeling of security suggested by their familiarity with the production documents and the manipulations fresh off the process start-up. On other words, *consolidation of the new manufacturing operation* is the foremost objective after technology transfer.

Next, the to do list should be executed promptly within the change control procedure. Measures for increasing production output should be conceived and taken as far as planning in the event of product growth. If the latter is already anticipated with some precision, those measures should be pursued deliberately.

If expedients with respect to raw materials were used to get to first manufacture (e.g., a single supplier of a critical material, a risky inventory position on another, etc.), those need to be addressed immediately, particularly if expanded output is desired. On the other hand, superfluous requests to qualify new suppliers (usually coming from Materials Management) should be rejected for the time being.

Process changes for cost reduction should be pursued on the basis of their technical soundness and merit first, then on the basis of their cost impact, but mindful of the possible introduction of too many changes too soon, as well as of the vicissitudes of supplemental submissions to the regulatory agencies. On the other hand, process changes aimed at increasing output, while scrutinized just as much, could be put on a track faster than those for cost reduction if firm production targets for the plant justify it.

In evaluating the above process changes, the original process body of knowledge should be mined with intensity, as invariably some good ideas and partially developed improvements have to be set aside during the original development if the dossier target dates are to be met.

For intermediates or bulk drugs made by contractors, the same approach embodied by the above recommendations should be sought in their operations, with particular attention to their observance of sound change control procedures and drug master file maintenance.

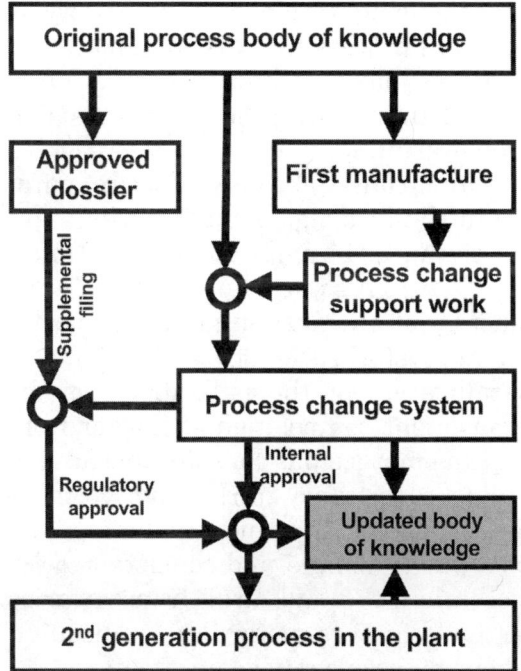

Figure 14   From first manufacture to mature manufacture: The process change mechanism that leads, over time and various process and operational changes, to a mature manufacturing operation that supports product growth and reduces the cost of goods.

Figure 14 attempts to depict the sequence towards mature manufacture just outlined in order of priority.

IV. IN CLOSING—THE PROCESSING TECHNOLOGIES OF BULK DRUGS

It is not in the scope of this chapter to address this topic in any breadth and least of all in any depth. The variety of the technologies is too large and the field far too rich. Instead, some selected observations that the new practitioner might find useful follow.

1. Process development organizations that lack a sufficient engineering component often miss the opportu-

nity of the better implementation of the chemistry by shunning continuous processing. It is not a matter of disdain, but of not having the tools and, often enough, the pull of the familiar batch or semibatch methods is too powerful.

2. Product purity and consistency, which are paramount norms in bulk drug manufacturing, are today observed through the impurity profiles to an unprecedented extent. This puts a great deal of pressure on the mastery of purification methods, mostly on those based on crystallization from solution.

3. A final recrystallization of the bulk drug for the purpose of a consistent composition of matter from which the material emerges has considerable advantages in providing consistency of the physicochemical attributes of the bulk drug, including the control of phase purity (single and consistent polymorph). It also buffers the bulk drug from vagaries upstream and tends to becalm regulatory disquiet, particularly about process changes upstream.

4. Scale-up of chemical processes is a business of much skill, largely because of the frequent intrusion of physical effects on the chemical kinetics. Good predictive tools and solutions exist, however, to deal with those intrusions by changing the physical environment away from that of the bench scale experience, but requiring the application of chemical engineering skills and the willingness to abandon the familiar batch or semibatch stirred tank when necessary or advantageous.

5. The bulk drug/dosage form boundary of process development is very difficult, often because of the discipline differences and just as often because the definitive decisions on the dosage form side come late in the cycle (often for good and largely unavoidable reasons). The bulk team must be sensitive and skilled in delivering to the dosage team what they need, and get very involved with their issues early

in the development cycle, particularly to seek multi-disciplinary decisions.

6. Practically all prescriptions for sound and successful bulk drug manufacture given in this chapter apply to the varied, seemingly tumultuous outsourcing environment, but only if a diligent effort goes into operating, maintaining, and building trust in the customer/supplier relationships.

In closing, few industrial endeavors offer as many opportunities for exciting and valuable technical work as the development of processes for bulk drugs and their implementation in performing chemical plants. The merging of chemistry (in its various fields), microbiology, chemical engineering, and pharmaceutics makes it possible, but demands that the practitioner of a discipline be earnest in the interaction with the others, regardless of their disciplines or their functions. Only through such effective interactions can success be reached in the exciting and difficult business of bulk drugs.

REFERENCE

1. Pollak P. From a commodities business to the world's leading manufacturer of exclusive fine chemicals. Chim Oggi 1997; 15: 75–81.

4

Design and Construction of Facilities

STEVEN MONGIARDO and EUGENE BOBROW

Merck & Co., Inc., Whitehouse Station, New Jersey, U.S.A.

I. INTRODUCTION

The design and construction of active pharmaceutical ingredient (API) facilities is an extremely complex and challenging undertaking. The time required to design construct and validate a facility to manufacture API products must be balanced against marketing and regulatory considerations. A firm may be required early in the drug development process to start investing in new production facilities or enhancing existing capacity so that a product can be produced for testing, and eventually for full-scale production to meet the market demand. An API manufacturer must develop a comprehensive process/facility design and construction execution strategy to ensure achievement of all regulatory, cost, and market objectives for the compound.

The successful completion of a new API process facility is a function of good engineering practices, sound construction techniques, and a well-planned and documented start-up and validation plan. Early detailed process definition enables the project team to develop a comprehensive project execution strategy. The execution strategy outlines the engineering and construction methods for the project. The start-up and validation plan ensures regulatory compliance and a smooth transition from construction to operation. Active pharmaceutical ingredient production facilities are complex, expensive to design, construct, and validate. New facilities require sophisticated processing equipment, utilities, and support functions. Careful planning and good sound engineering is critical to assure that the investment in capital is managed wisely.

The reader must be cognizant of current good manufacturing practices (cGMPs) requirements for new products. The design engineer will be responsible for design of facilities and systems that will meet cGMPs for API manufacturing. The constructor will be required to install and validate the equipment and facilities to meet the same criteria. Certain utilities, such as process-deionized water, are required to meet specific regulated criteria. Details of validation and cGMPs are discussed in other sections of the manual. We will focus on the impact to design and construction by validation and cGMPs.

Various strategies utilized in the engineering and constructions of new production capabilities, whether they are new facilities or renovations to existing capacities, are reviewed. A clear execution strategy and the need to understand the scope of the project are important components that are reviewed in detail. Management of the design process is critical to success. The firm must be able to properly manage the development of the design, ensuring that the process is complete and "frozen" prior to the commencement of construction.

Many of the terms and references in this section are typical of fine chemical manufacturing. We will assume that the reader has an understanding of fine chemical manufacturing.

II. BUSINESS REQUIREMENTS

An API project is created when a need arises through one or more of several areas. Some examples include: (1) new product introduction, (2) regulatory requirements, (3) existing product capacity shortfalls, and (4) process improvements. Any one or combination of these areas can generate the need for a new capital investment.

The engineering and construction steps are similar for the four stated cases. Excellent scope definition and a well thought out execution strategy are required for all of them. A firm will analyze many different manufacturing options before establishing the final project scope. The most uncertainty occurs with the new product introduction. The firm can be required to develop a preliminary scope of work during early stages of process development. There is a higher probability of change and process churn as the new process develops. The firm must be prepared to manage facility and equipment changes as the new process is finalized. The key to success is both minimizing and coping with those changes. The uncertainty in the volume of production requirements can also change during the initial scope development of a project.

The other business cases normally have a defined process within existing operating facilities and with known market volumes having already been on the market. The major process components have already been defined. The process is a regulatory agency-approved process, which has proven viability. Normally, the changes associated with these types of projects are limited in scope to process enhancements, i.e., increasing throughput, eliminating bottlenecks, increasing yields, etc. We will not focus the discussion on these types of projects.

We will focus on the requirements of engineering and construction of facilities for a new API introduction. This is the most difficult and complex task because the technology is untested on a commercial scale and there are technical assumptions with the associated risks that must be taken. Assuming the product and the process are untried at commercial scale, there may be unforeseen issues with start-up and operation that arise at the proposed scale for the new API entity. Certain components of the process, such as product handling and transfer, and material consistency may become an issue at the production scale, which were not detected in a pilot or bench operation. The design team must take into consideration any components of the process that will not be a scaled-up duplication of the laboratory version of the process. The risk of a pump or product transfer system not working properly because of material viscosity or incompatibility may require changes to the process once the system is built.

Material handling aspects through equipment such as centrifuges, blenders, or mills can be different from the smaller-scale experience. Common problems that develop at commercial scales include pumps not operating as designed, material bridging in centrifuges and blenders/dryers, and different milling consistencies. The engineering and construction team may be required to change components during the initial production runs of a new compound. A good designer will incorporate the necessary flexibility in the new process to allow for equipment change outs. The design of equipment should incorporate the ability to replace it or upgrade in a manageable fashion.

A benefit for the readers is the ability to utilize this strategy for other process improvement or regulatory-driven projects. The steps are similar if not the same (the major differences are associated with the business and engineering analyses for the new API).

A. An API Manufacturer Will Focus on the Appropriate Level of New Capacity

Market projections will indicate required volumes of the new API. Unfortunately, market projections can vary widely. Manufacturing capacity for a new API facility is expensive to build, maintain, and operate. It is important to "properly size" the production processing equipment and supporting facilities. A proper engineering approach will incorporate the ability to expand a production facility or equipment train(s) to allow for future expansion for potential volume increases. The future expansion planning can be as simple as incorporating the footings for a building expansion during construction of a new facility. It can be as complex as adding an additional bay to a new or existing building along with all associated utilities for that future expansion.

B. The New API Manufacturer Will Focus on Flexibility in Design

A key component of the analysis is whether to produce the new API with dedicated process trains and facilities or to *campaign* the new product with other products utilizing similar equipment. Major new API compounds may warrant dedicated process trains and facilities due to the sheer volume of product or due to unique processing techniques. However, most new higher-potency APIs can be produced with less equipment over a shorter time span than in the past, thus allowing the manufacturer to produce multiple products within the same equipment. This approach can often result in significant conservation of capital.

C. Location of the New Facility

The new facility can be located in any of the major markets in the world. Many countries provide tax incentives for locating an API production facility in their country. Labor markets are an important component of the analysis. The technical skills required to operate and maintain the facility and for construction and start-up are sophisticated. Complex processes require skilled technicians to run the process. Skilled mechanics will be required to maintain the facility.

Sophisticated equipment and facilities require skilled labor and construction professionals. It is difficult to construct and maintain one of these facilities in a remote part of the world and certain parts of the United States. Labor markets are limited. The new facility may compete with other facilities under construction for the available labor and construction support resources. Major API and biotechnology projects have recently experienced large cost impacts, both in the United States and foreign locations, because of a dearth of trained construction and engineering professionals to design and construct these facilities. Site selection should be one that offers an acceptable supply of operational support and construction resources.

III. DEVELOPING THE PRELIMINARY SCOPE

A preliminary scope of the new process should be developed in parallel with process chemistry and engineering development and early piloting for the new API. The scope should include a definition of the process, preliminary process flow diagrams (PFDs) and piping and instrumentation diagrams (P&IDs), equipment specifications and requirements (vessel types and sizes), preliminary facility fit, permitting requirements (local building and environmental), and any regulatory (cGMP) requirements. The producer's process-engineering group will have to determine the best process fit to ensure speed to market, cost-effective manufacturing, compliance with safety and environmental requirements, and GMP compliance.

An analysis of alternatives is desirable once enough process definition is developed. Process siting and development decisions should be based on scientific, business, and regulatory analysis. Questions the producer should consider in the preliminary scoping exercise include:

- Should the facility be multiuse (Fig. 1) (campaigning) or dedicated?
- Can the API be manufactured in an existing facility (retrofit) or will it require a new facility?
- Will the new/retrofitted facility be cGMP compliant?
- What are the safety and environmental concerns of the new compound?
- Utilities:
    - What is the status of process water and building utility systems?
    - Do existing assets have the utility capacity(s) for expansion?
- Schedule and cost: What is the best approach to support a product launch?

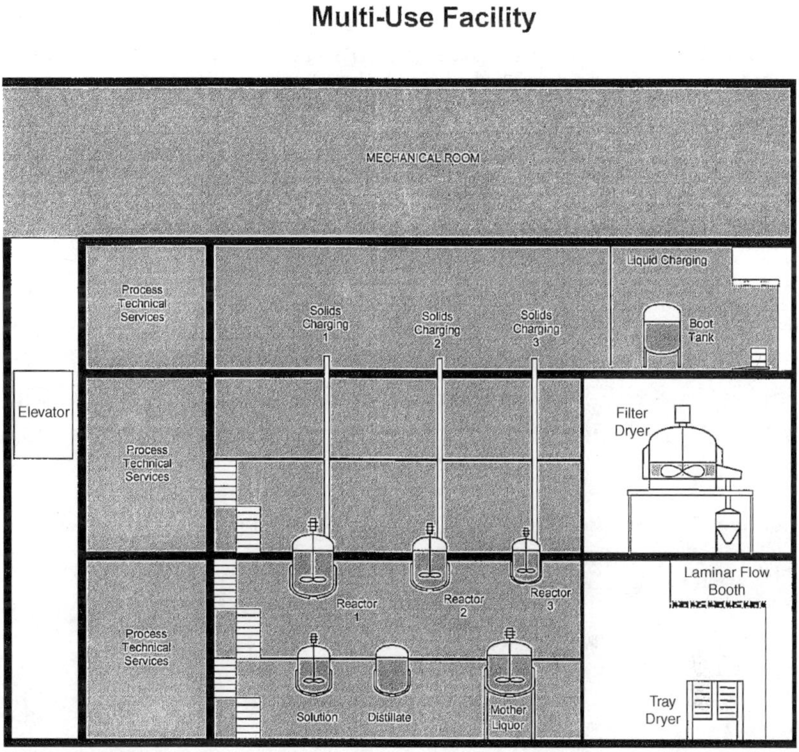

Multi-Use Facility

Figure 1 Typical vertical processing facility designed for batch production with barrier separation between processing steps.

A. Campaign vs. Dedicated

Many of the newer compounds developed for market are of a higher potency, reducing the need for large (greater than 1 million kilograms annually) volumes of the API. A dedicated process may be the easiest approach to design and construct, but may not be the most cost effective or strategic. A dedicated process is ideal for a one-product organization or high-volume product. It may be easier to manage, with unchanging processing parameters. Varying market product demand can impact usage of the facility and the cost of operations.

A campaign style facility will allow the manufacturer to better utilize assets, integrating different product manufacturing using similar equipment configurations. The producer has different options available for product volumes and production time. The campaigning facility will have different processing capabilities (Fig. 2) through various manifolds or hard piped equipment configurations (equipment trains). These configurations can be manipulated for different processes. This provides the manufacturer with the flexibility to vary production sequencing to produce several products vs. one.

Good manufacturing practice (GMP) considerations must be reviewed carefully with a multiuse facility. Good manufacturing practices controls are applied with the use of API starting materials. The controls increase as process proceeds to final isolation and purification. The producer will

Chemical Processing Spec. Vent for Reactor Charging

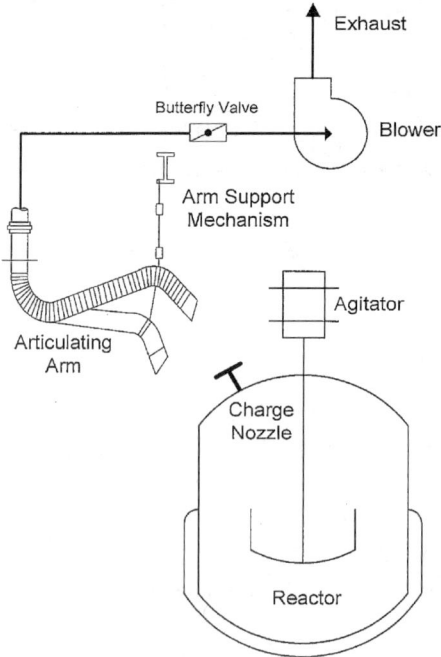

Figure 2 Typical specific ventilation system designed to protect operators during the care of vessels with cytotopic-high potency compounds.

be required to ensure GMP integrity for the new or renovated facility. Some of the considerations include proper product isolation (barrier separation), cleaning systems for multiproduct equipment (CIP—clean in place) and pharmaceutical grade water systems (for isolation and purification).

B. New vs. Retrofit

The API manufacturer will be required to decide whether a new facility will be required or an existing facility can be retrofitted. Questions that the manufacturer should ask include:

- What existing assets are available in the manufacturer's portfolio?
- Can these assets be modified to process the new product?
- What are the costs associated with the renovations? How do they compare to a new facility(s)?
- What renovations are required to qualify the process or facility?

A careful cGMP review will be necessary as part of analyzing an existing facility for a new product fit. The current guide for GMP guidance for API facilities is the *International Conference on Harmonization (ICH) of Technical Requirements for Registration of Pharmaceuticals for Human Use guide Q7A*—referred to as ICH Q7A (1).

Facilities currently manufacturing fine chemicals may not meet the standards outlined in ICH Q7A and not hold up under the scrutiny of a regulatory inspection. Major renovations may be required retrofitting an existing facility(s) to assure GMP compliance. The manufacturer can be required to install new systems such as CIP and pharmaceutical grade process water. In product isolation/purification and finishing facilities, the manufacturer will have to insure product separation in multiproduct suites, through physical barriers such as walls, and also through differential room pressurization with minimum room air exchanges.

Understanding the requirements of a GMP facility is critical to developing an accurate cost and schedule model for the new product. A process fit that appears simple for a fine chemical could require substantial renovations for an API.

C. Equipment and Facility GMP Compliance

Good manufacturing practice regulations affect the architectural and building engineering components of the building along with equipment and systems. The building must be capable of providing items such as adequate lighting, proper waste water management, validated process water, product separation areas (warehousing), and heating ventilation air conditioning (HVAC) and room separations for final step (isolation/purification) processing. The facility should have the appearance of a pharmaceutical facility. The processing areas should be clean and free of debris. "Cleanability" is critical for all processing equipment involved in "critical step" and post-critical step manufacturing.

D. Safety and Environmental Concerns

Many of the new APIs are designed with a higher potency (cytotoxic) than previous generations. The stronger potencies require the designer to integrate materials handling and HVAC systems that protect the operators from exposure to the product. Specific ventilation systems are incorporated to protect personnel while charging and operating vessels (Fig. 2). The facilities are designed to contain all materials within the confines of the facility. Similar to sterile processing, there will be air locks separating the different rooms (Fig. 3).

High potency facilities will normally have separate compartments for gowning/dressing and entering, processing/manufacturing, and decontamination/degowning. The HVAC system will be dedicated for the facility. Wastewater will be discharged to a holding tank for testing prior to disposal. The concept for the facility is total containment.

Process wastes will be managed similar to any organic fine chemical operation. The producer must separate and contain all waste materials not suitable for wastewater treatment.

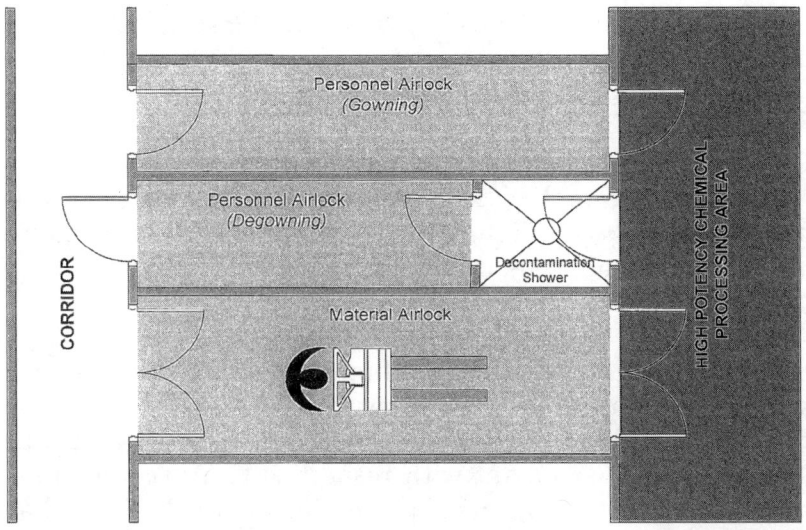

Figure 3 This high potency materials all lock provides separation of materials from personnel and provides separate pressurized entrance and exit points. The high potency processing space will be negatively pressurized to the adjoining spaces effectively containing any airborne materials exposed during processing. Operators enter through the air lock, change into protective gear, and enter the processing area. The doors are typically interlocked—not allowing someone to enter into the processing area if the outer door is open. Once work assignments are completed, individuals exit through the degowning chamber, showering and removing contaminated outerwear before exiting. The shower water is contained in a holding tank for disposal. The unit is fully self-contained.

IV.   UTILITIES AND BUILDING SYSTEMS

A.   Process Water Systems

Water systems are expected to be *demonstrated* to be suitable for their intended uses. At a minimum, water is required to meet the World Health Organization requirements for drinking (potable) water. Processing steps such as isolation and purification will require purified water as outlined in USP 23 (2), pharmaceutical grade water. Validation of water systems is required for all product contact water systems. We will discuss design of purified water systems in more detail later in this chapter.

B. Gases

Gases used in final processing steps will also require validation and GMP compliance. These gases, such as nitrogen, will be required to pass through filtration systems to remove any microbes that might be in the gas stream.

C. Heating Ventilation and Air Conditioning

During development of the preliminary scope, the engineer should take into consideration any HVAC control issues for the new product. The design of the new or retrofitted facility must be cGMP compliant with respect to HVAC controls for all "final step/post final step" processing areas. Products that have specific temperature or humidity requirements must be manufactured in facilities that will assure regulators of those conditions for critical and postcritical processes. Dry processing steps such as milling, drying, and blending are to be performed in areas that assure the manufacturer and regulators of no product cross-contamination.

V. PRELIMINARY SCOPE DELIVERABLES

The preliminary scope should include enough information for the engineer and constructors to start developing cost and schedule data. The information should include process flow diagrams, preliminary P&IDs (Fig. 4), initial facility requirements, and the first cut at a validation strategy.

A. Contracting Strategy

Once the API manufacturer has generated enough initial information for the new process, they will have several options for implementation of the design and construction of the new facility. There are various execution strategies, which include:

- design and construction utilizing an engineering firm and a construction contractor(s);

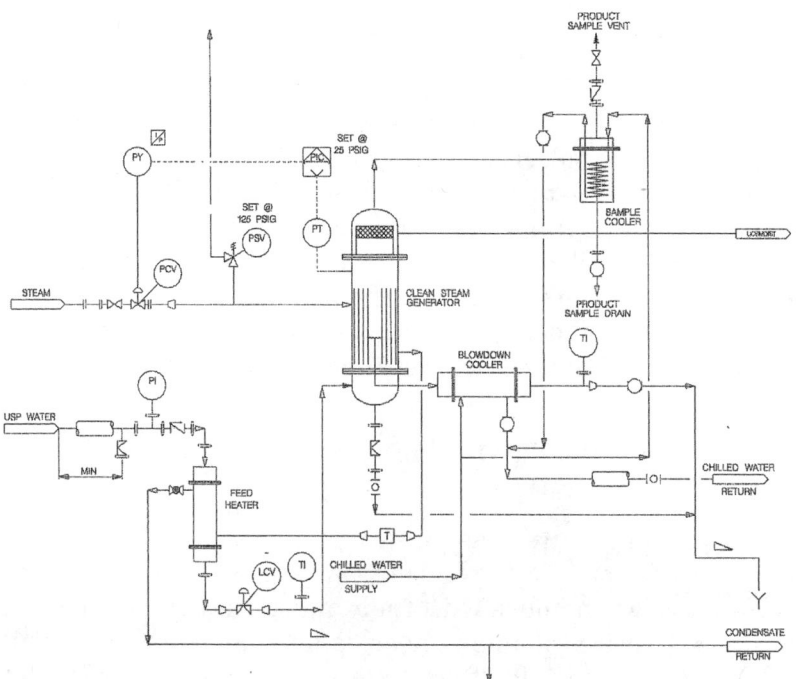

Figure 4 Example of a P & ID for process utilities system (clean steam generation).

- design and construction utilizing an engineering firm(s) and a construction management firm(s);
- utilizing one firm to provide engineering, construction, and procurement services.

The use of the one firm concept of engineering, procurement, and construction is well suited for a manufacturing firm that does not have depth within its own engineering and procurement organizations. This method places all the responsibility on the engineering contractor to deliver a finished GMP compliant facility. The firm selected must have the personnel depth to be able to supply all facets of the project. This method tends to be the most effective for schedule, but can carry cost premiums. The engineering, procurement, and construction (EPC) contractor assumes all the risks on the project and will charge a premium for assuming the risk.

The other methods work better with manufacturing firms that have established plant and construction engineering groups. The first method, design and contract, is the "traditional" method of construction. The design is completed and the project bid. The successful bidder has a lump sum contract to complete the work. This method tends to take longer. However, in a competitive market it can be the most cost effective. Utilizing a construction manager and separate design, firm will enhance the schedule of the project by bidding work as it is designed and will control costs if properly executed. The chapter discusses these methods in detail later.

B. Development of the Design Strategy and Detailed Design

The API manufacturer has essentially two choices in setting a design execution strategy. They may elect to develop the detailed design in house with their own expertise or they will obtain the services of an engineering firm/contractor who will provide the services for them. We will discuss the option of utilizing outside services for this function. The vast majority of manufacturing firms do not possess the "in-house" capabilities to develop the full breath of design for a new process facility or a major renovation.

There are various methods of employing the outside firm utilizing various contracting strategies. The firms can be hired on a reimbursable basis. This method is the most common in the industry. The firm is remunerated for all design services cost plus a mark-up for overhead and profit. Typically, the manufacturer will negotiate a contract with the design services supplier for a "not to exceed" value for the work. The design firm normally develops this estimate. It is impacted by the stage of process development. The more defined the process and the scope of work, the better the estimate.

The other less common method is buying the design on a lump sum basis. The design firm provides a firm price for the work. A careful definition of expectations is required for this approach. This contracting strategy is akin to using a building contractor for a home or commercial building. The pricing is based on a fixed set of parameters, which are normally the plans and specifications developed for project. Any items not included in the plans and specifications are considered out of scope. All items out of the scope of the contract are subject to extra charges. The lump sum design contracting strategy is a difficult strategy for designing a new process with uncertainty. As design progresses, any changes to the process will result in negotiating change orders to the contract with the design firm. This method can create distractions to the design effort as the manufacturer and the engineer become involved in pricing negotiations. This method is more common with small process configuration changes in existing facilities.

Selecting the right firm and establishing clear expectations is critical to the success of the project. How the firm is utilized is a decision to be made in the planning stages of the project. The API manufacturer can elect to use the various methods of contracting for services that have been outlined in the chapter.

C. Setting Expectations

The execution of the design and construction process for an API facility can be defined in four critical steps:

1. design
2. procurement and construction
3. equipment validation
4. start-up, commissioning, and turnover

We have briefly discussed different contracting options. The API manufacturer has to decide how to procure the outside services necessary to accomplish these steps. We cannot recommend any one method as better than another. The contracting decisions must be made based on all party's relative strengths in the execution of this type of project. If the decision is to contract out the entire process to one firm in an EPC contract, the outside firm is expected to deliver a completed facility, validated and ready to produce product.

The API manufacturer, with some level of in-house expertise, can elect to manage the design and construction separately. The outside design firm will be responsible for providing the proper level of documentation for a construction firm to execute the work. The API manufacturer exercises a greater level of control in this process. The API manufacturer will be involved in many of the decisions made in procurement of equipment and other components and be better able to influence the operability of the facility.

VI. DESIGN DEVELOPMENT

The design of a new API facility will develop from an initial "napkin" exercise to a set of documents that a constructor will use to install the new assets. Progression of the design can be inferred from the following sequences: preliminary scope, basis of design, and detailed design. Many companies in the petrochemical and chemical industries utilize this practice. Recently, this progression has been utilized in the API industry.

Development of the design is a function of process definition. Once the process is clearly identified, the API manufacturer and/or the design firm can complete the in-depth analysis of existing assets and new assets to progress the design for the project.

As previously discussed, the preliminary scope defines the major components of the API facility. The scope will have identified key processing steps, all associated equipment, and any facility requirements. The preliminary scope will be a key document in communicating to outside design firms the intent of the facility(s) and the overall process intent of the project. The API manufacturer must then decide how they want to manage critical steps of the design, construction, validation, and start-up process.

Another phase in design development currently utilized in the chemical/petrochemical industry is a "basis of design" phase. The basis of design has certain components of the design defined prior to a final cost estimate and schedule is completed. The basis of design has components such as "approved for design" P&IDs, PFDs, and facilities definitions already determined. The basis of design will outline: permitting requirements—local government and environmental, engineering criteria for the new site (civil and infrastructure), and a preliminary project execution strategy. Typically, the basis of design represents approximately 20% of the total design.

Once the basis of design is completed and the final estimate generated, the design team will develop detailed design documents, which will incorporate all the necessary information for the builder(s) to construct, and start-up the facility(s). The following components should be defined during this process:

- equipment requirements
- facilities requirements
- utilities requirements
- safety requirements
- cGMP requirements
- qualification or validation plan
- expansion capabilities
- hazard and operability (HAZOP) analysis
- process and instrumentation diagrams (P&IDs)
- enviornmental requirements (permitting)

A. Equipment

The P&IDs will identify the major equipment components to be modified or procured for the new process. It will outline the new equipment and associated controls required to run the process. The API manufacturer may have preferred suppliers of this equipment because of operability or maintenance issues. Most of the vessel and component suppliers in the industry are capable of supplying cGMP compliant equipment. The design firm or the API manufacturer will specify the equipment to include the necessary appurtenances to make the equipment cleanable.

The design firm normally generates procurement specifications. These specifications will include definition of all major components such as materials of construction, agitator requirements, nozzles, etc. Major equipment for API processing is similar to fine chemical production and can include reactors, centrifuges, condensers, heat exchangers, distillation columns, extractors, absorption equipment, chromatography equipment, dryers, blenders, crystallizers, mills, etc. These components will normally require validation [installation qualification (IQ)/operational qualification (OQ)/performance qualifications (PQ)] prior to manufacturing. It is important to have proper coordination between the design engineer and the validation team. The validation team will be responsible for generating validation protocols developed from the equipment specifications and processing parameters.

B. Facilities

Many of the recent cGMP initiatives have been focused on facility requirements. The manufacturer will be responsible for adhering to these requirements for processing, product separation, materials handling, and utilities. Because of these requirements, additional space for functions such as warehousing may be required. ICH Q7A is the document the producer will refer to for information on facilities and processing requirements. The new facility will require enough space to provide separation of raw materials (i.e., quarantined vs. approved) and for finished products and intermediates (quarantined vs. approved). The manufacturer must be able to isolate and segregate these components.

Manufacturing space, containing equipment involved in critical and postcritical step processing, is required to maintain a level of cGMP compliance consistent with the stage of the product. There are noticeable differences between intermediate facilities and facilities that manage final (critical and postcritical) steps of an API process.

An intermediate facility will not require the level of sophistication normally associated with a final process facility. There will be differences with respect to equipment separation, architectural finishes, utilities, and general building configuration. Some of the differences include less costly equipment separation (open bays vs. rooms); architectural finishes consistent with fine chemical manufacturing, and potable (drinking) water vs. purified water.

The isolation/purification facilities will be designed with GMP considerations consistent for the final stage of the product. Some of those considerations include product isolation and separation (Fig. 5), cleanable surfaces, purified water systems, and temperature, airflow, and humidity controls.

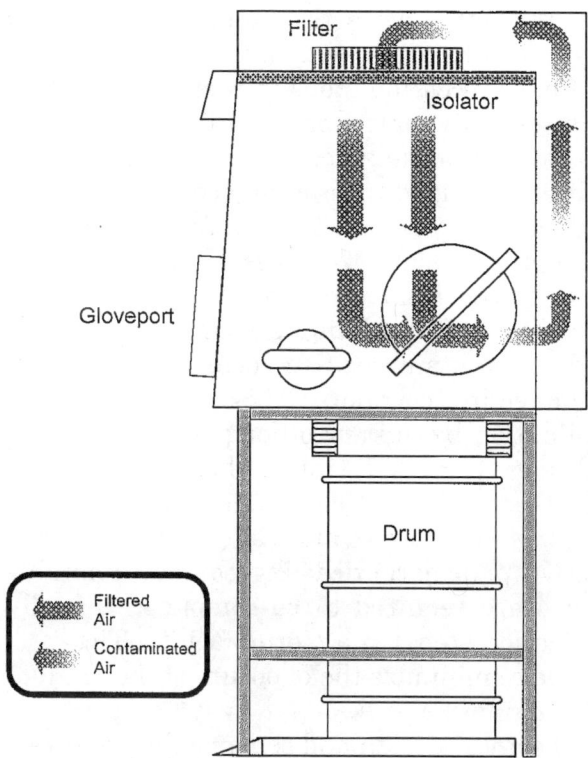

Figure 5 This figure represents a self contained product transfer "box" that allows product to be transferred from a piece of equipment such as a centrifuge to a drum protecting the product from contamination. The unit also provides operator protection from high potency compounds.

C. Product Separation

Product isolation is required for processing facilities involved in critical and postcritical processing. Production facilities involved in campaigning different products must be able to provide adequate separation for product isolation during transfer of product. The producer must be able to assure the regulatory agency that all products are produced in an environment free of the potential of cross-contamination. This isolation can be performed through the use of temporary barriers such as curtains or through the construction of separate bays with permanent physical barriers (walls).

Isolation for final processing steps is best accomplished through the design of individual rooms (suites). The rooms are finished with smooth, cleanable, durable finishes such as epoxy. All utilities are piped in from adjacent mechanical rooms. All penetrations are sealed to maintain the proper environment in the process suite. The vessel heads and man ways should be kept free from overhead components that could collect dust. Heating ventilating air conditioning systems are designed to maintain positive room pressurization with respect to connecting corridors. This will protect the room from contamination by dust or other particulates. The HVAC system will have particle filtration that will filter out airborne contaminates.

Sterile processing facilities require additional levels of sophistication. Active pharmaceutical ingredients manufactured for sterile use are required to be completed (usually the isolation/purification steps) in a sterile facility. The sterile facility is designed to minimize the exposure of the product from microbial contamination.

CFR 211.42 (3) states: (design and construction features) requires in part, that aseptic processing operations be "performed within specifically defined areas of adequate size. There shall be separate or defined areas for the firm's operations to prevent contamination or mix-ups." Aseptic processing operations must also "include, as appropriate, an air supply filtered through high efficiency particulate air (HEPA filters) under positive pressure," as well as systems for "monitoring environmental conditions..." and "maintaining any equipment used to control aseptic conditions."

Section 211.46 (ventilation, air filtration, air heating, and cooling) states, in part, that "equipment for adequate control over air pressure, microorganisms, dust, humidity, and temperature shall be provided when appropriate for the manufacture, processing, packing or holding of a drug product." This regulation also states "air filtration systems, including pre-filters and particulate matter air filters, shall be used when appropriate on air supplies to production areas."

The building or suite is designed with separation and control. Air quality will vary depending on the nature of the operation. The area design is based upon satisfying microbiological and particulate standards as defined by the equipment, components, and products exposed, as well as the particular operation conducted, in the given area.

There are two clean areas that are important to sterile API product quality. The critical area (class 100) and the supporting clean areas associated with it.

Class 100 conditions require that air in the immediate proximity of exposed product be of an acceptable quality with a particle count of no more than 100 0.5-μm particles per cubic foot of air. This is obtained by utilizing HEPA filters and laminar flow conditions with the room HVAC. Room pressurization is also critical. Class 100 rooms are required to maintain a positive pressure to surrounding rooms of at least 0.05 in of water will the doors closed. The supporting clean rooms can vary from class 1000 to class 100,000 depending on the function with class 100,000 as the least critical to class 10,000 for adjoining rooms.

The sterile processing buildings are designed with a hierarchy of separations. Manufacturing is separate from warehousing, warehousing and manufacturing from offices and locker facilities, and also separations from utilities. Material flow is critical for a successful design of aseptic facilities. The designer should make every attempt to design the facility for unidirectional flow of components, ingredients, and product. Unidirectional flow means that materials and product all flow in one direction as the product steps through various phases of completion. This provides the greatest assurance that sterility will not be compromised as it might in a facility that did not possess unidirectional flow.

VII. UTILITIES

Utility requirements for API manufacturing are not atypical from fine chemical manufacturing. Process equipment will require steam, water (potable, chilled, tower/cooling), nitrogen, and vacuum. Jacket services can be designed for multiple fluid use or for single fluid applications, utilizing a multipurpose fluid for both heating and cooling. Buildings should be designed with utilities separated from the processing areas. Chillers, heat exchangers, pumps, etc. should be located in separate rooms, floors, or areas.

A. Water Systems

Water systems used in the manufacturing of APIs are subject to validation guidelines as outlined elsewhere in this book. The manufacturer will be required to utilize a water system as appropriate to the process. The United States Pharmacopoeia outlines minimum specifications for various levels of water purity. The manufacturer will determine the quality level required at all the various stages of molecule development. The manufacturer will utilize drinking water quality, potable water, for all precritical processes (intermediate steps) and purified water for critical and postcritical applications.

Purified water systems (Fig. 6) for pharmaceutical processing require a level of sophistication not required in the production of fine chemicals. There are many different ways to generate purified water. Following is a brief description of a typical pharmaceutical grade water system. The system will include the following components:

1. treatment
2. sanitization
3. storage and circulation

1. Treatment

The equipment required for water treatment will be determined by the quality of the incoming water. Typically, a USP pharmaceutical grade system will require pretreatment (filters), deionization, reverse osmosis, and potentially a polishing step such as continuous deionization. Many systems now incorporate UV filters for sanitization, which kill microbials and also eliminate ozone.

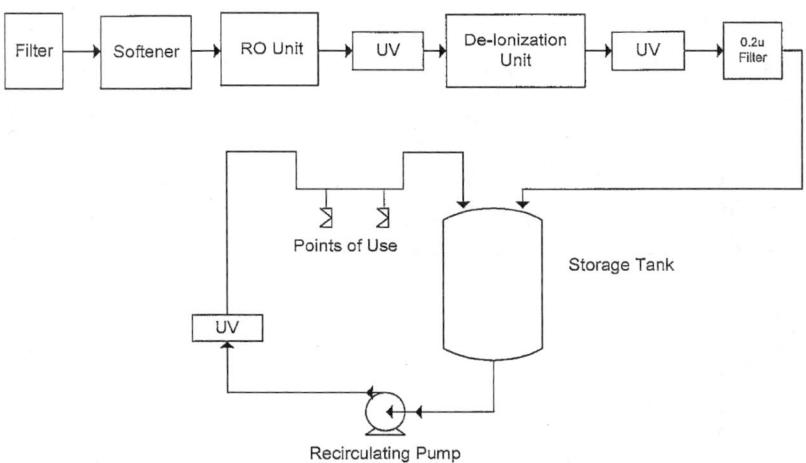

Figure 6 Example of a circulating purified water loop utilizing reverse osmosis, deionization, ultra violet lighting, and micro filtration.

2. Sanitization

The systems are designed to be cleaned. Recent industry practices have included the use of ozone injected into the system as a sanitization step. Other methods include steam sanitization. The ozone is then eliminated through the UV filters. The system should be designed for a complete sanitization, which includes all storage tanks and distribution piping.

3. Storage and Circulation

Typical systems include a storage tank with sanitization capabilities. Treated water is sent to the tank and then circulated to the points of use in the manufacturing areas and returned in a continuous loop. The circulation loops are normally designed to maintain flows that will inhibit bacterial growth within the distribution systems. The system designer must be aware of the minimum velocity requirements for these systems. Typically, the designer will use a rule of 4–6 fps (feet per second) as a minimum velocity.

4.   Materials of Construction

The system will be constructed of sanitary materials. All fittings will be sanitary grade. Typically, these systems are built utilizing PVC or stainless steel (316L), or a combination of the two.

Other systems impacted by cGMP requirements (product contact) include nitrogen and plant air. These systems will require filtration systems to insure no impurities are passed through and make contact with product.

VIII.   SAFETY

As previously stated, process and facilities designs are impacted by the potential need to include handling of hazardous (cytotoxic) compounds. The newer higher potency compounds are potentially toxic in the large volumes they are produced.

Heating ventilation air conditioning systems for finishing facilities where potent/toxic compounds are handled are required to work under negative pressure. High efficiency particulate air or 95% air filtration systems are utilized to remove particulate from the air stream. In intermediate processing, isolation chambers may be required to protect the facility environment. These chambers will contain the equipment that holds the material (i.e., centrifuge or dryer) and provide a physical barrier (plaster or block walls) and an air bath (under negative pressure). Operators working in this environment will be required to wear personal protective equipment (PPE). The facilities will be designed to minimize the hazard by limiting exposure to the individuals and environment.

IX.   CURRENT GOOD MANUFACTURING
PRACTICES REQUIREMENTS

Current good manufacturing practices requirements have been discussed throughout this section. The reader must be aware of the requirements of ICH Q7A—good manufacturing practices for API facilities. This document provides guidelines for manufacturing facilities for API products from introduction of the API starting materials through physical processing and packaging.

X. QUALIFICATION PLAN

The qualification plan for an API facility consists of the following:

- commissioning
- validation
- installation qualifications
- operational qualifications
- performance qualifications

The supplier of engineering services can also provide commissioning and validation services for the API manufacturer. There are also third parties specializing in commissioning, validation, standard operating procedure (SOP) writing, and operational training. Early in the project development process (initial scope development) the contracting strategy for validation/start-up services should be determined. Engineering, procurement, and construction contracts can include this as part of the suppliers' scope. The API manufacturer will have to decide how this work will be executed. The API firm can perform the work internally with engineering resources. They can use the engineering firm of record or the contractor to perform the service—or they can hire a third party specializing in validation/commissioning.

The engineering firm will be responsible for identifying key supplier documents required for validation (IQ and OQ) and also specifications and ranges for equipment. The validation services will develop protocols for executing each component (IQ and OQ). Performance qualifications are performed after completion of OQ. The engineer or contractor will typically not be involved in PQ. Our experience has been that the producer will perform PQ on the new process.

The IQ protocols are designed to verify that the installation has been completed as specified. As an example, an IQ protocol for a vacuum pump will ensure that the right pump was installed, as specified. The entire nameplate data will be recorded, documenting all necessary engineering information such as size, type, and purpose. All electrical and instrumentation contacts will be tested and verified.

The OQ protocols will test all critical parameters for the equipment. It will test all control devices, calibrate critical instruments, and test major vessels under operating conditions (pressure and vacuum).

XI. EXPANSION CAPABILITIES

Earlier in this chapter, we identified the importance of "correctly sizing" the process. Market projections for API products can swing wildly. A recommendation for the reader is to be cognizant of the potential for expansion. When designing the new process facility or upgrading an existing plant, the API producer should position the plant to be expanded if necessary. Consideration should be made for preinvestment of some facilities or utilities during the design of the first phase. It is clearly cost effective as compared to having to reinvest later. However, it is ultimately a management decision on managing risk. As previously mentioned, certain facility components are easily installed at one time vs. staggered. A good example is piles and foundations. If the producer anticipates product growth, the scope may include the necessary subsurface components for future expansion. When the producer decides to expand some time in the future they will be able to avoid costly foundation excavations and associated disruptions of heavy subsurface civil construction.

XII. HAZARD AND OPERABILITY ANALYSIS

At various stages of the design process, a HAZOP (Fig. 7) must be conducted on the project. The purpose of this analysis

Sample HazOp Report

Deviation	Causes	Consequences	Safeguards	Recommendations	Assigned to	Comments	Comment date	Drawings
No/Less Flow	Manual valve closed	Loss of N2 inertion resulting in potential fire or explosive atmosphere	SOP will contain instructions to guarantee proper valves open	None				
No/Less Flow	Flexible hose disconnected	Potential asphyxiation hazard	Solution prep room ventilation is 100% exhaust	Ensure O2 detection/LEL alarms from granulator room is activated also in solution prep room.	John Beam	Explore the possibility of using the exhaust flow from the room as an alarm to protect from asphyxiation scenarios. Provide a local, visual and audible alarms.		
Reverse Flow	Manifold valve malfunction	Potential to backup CIP/rinse and/or instrument air into N2 system	Limit switches on the valves; Minimum and drainable deadlegs	Check that N2 isolation valve is always exposed to CIP solution. If so, consider use of check valve to the N2 line.	Dee Chow	COMPLETE. N2 Isolation valve is not always exposed to CIP solution.		
Part Of Charge of solid excipients	Agitator rotating in wrong direction	Liquid pumping back up through charging tube resulting in the wrong recipe and potential spill	IQ setup to verify proper rotation of agitator	Ensure PM addresses check of proper rotation of agitator motor.	Tech Ops			
No/Less Flow	Bottom valve closed	Pump runs dry resulting in potential equipment damage	Limit switches; Process command disagree alarm	None				
Reverse Flow	Pump wiring incorrectly	Potential equipment damage to pump	IQ and PM checks	None				
Higher Temperature	Chilled water control valve connected to level switch	Loss of cooling	None	Clarify schematics for cooling water temperature control.	Dee Chow			

Figure 7 This sample HAZOP report identifies several hypothetical operating hazards that can occur during processing/operations. The report includes recommendations and safeguards assigned to specific individuals for execution.

is to identify any potential weakness in the design of a process facility. Weaknesses are identified as:

- safety concerns (i.e., dangers to personnel)
- enviornmental impact (i.e., chemical release)
- economic impact (i.e., damage or loss of equipment or facility)

The HAZOP reviews will look at each detail of the process, examine what is happening in that stage of the process, and then question a series of "what if" potential failures. Questions such as a failure of a control, loss of power, will generate a list of possible reactions to that failure mode. The failure list is then generated from experience with similar process arrangements or from experience with this specific equipment. This potential failure is then analyzed and a determination of the risks. A failure of low risk from safety exposure or cost of damage to the facility would not generate further action by the design team. An item with a potential of extremely unsafe condition or high cost damage would then be listed with a recommendation for additional controls or a revision to the design.

XIII. EXECUTION STRATEGY AND PLANNING

In any project, the cost and schedule predictability is important. All projects, from an office building to a major manufacturing facility, have external factors, which are affected by the cost and completion of a project. When a project is first conceived, the drivers for the project establish the cost and schedule goals. If a project is purely financially driven then cost is the major controlling factor. If speed to the market drives the project, then schedule becomes of paramount importance. In general, the project execution strategy must have a primary goal. If the project is falling behind schedule a decision must be made to determine if the cost of overtime is justified. If cost is the driver then a slip in schedule may be acceptable or, conversely, the cost of working overtime may be justified to maintain a critical schedule.

The "*project execution strategy*" must be aligned with the overall project goals. To align these goals we must first understand how executing a project can support these goals. If cost is the main goal of a project, such as in commercial or government projects, the contracting strategy is to have competitive lump sum bids submitted by as many general contractors as can demonstrate the financial strength to complete the project. This financial strength is demonstrated through the submission of a bid bond with the proposal. The quality of the end product is "industry standard."

This method of lump sum bidding requires all the design documents to be completed and the entire project construction is awarded to one general contractor. This contractor will be responsible for the procurement of the individual trade subcontracts and the procurement of all the equipment. Because the design is completed, and the basic coordination of the trades has been engineered in advance of the start of construction, the extent of extras on the project is usually limited to unknowns. Items such as subsurface conditions or, in the case of renovations, existing conditions, which are not anticipated in the design, may be uncovered during the construction activities.

The administration/management of a lump sum general contract is also limited due to the completion of the design prior to the start of the construction activities. This project approach will take the most overall schedule time, but the final cost is easily predictable and known with great accuracy prior to the start of construction. The delivery of long lead equipment and fabricated materials such as structural steel will often not support the overall project schedule. There will be times in the construction schedule when the project is waiting for materials and equipment to be delivered. In the construction of most API facilities, schedule is an important driver of a project. The lump sum general contract approach will not normally satisfy the schedule requirements of the project.

Schedule time is normally reduced through a technique known as "fast tracking." Major equipment bid packages and early construction packages (foundations, structural steel, etc.) are awarded prior to completion of the entire design. As the design progresses, other packages are released for bidding. A major component of the *"project execution strategy"* is the bidding strategy for the project. The system of fast track is normally applied to the entire project execution strategy. Almost all API manufacturing projects are schedule driven.

Cost predictability for fast track projects is less definitive. The potential exists for early construction packages to have errors or omissions that would have been uncovered if the remaining design had been complete. Whenever the design is not complete there is cost uncertainty. Any late changes can impact design components that have already been ordered or constructed. Accordingly, the administration of a fast track project requires extensive field effort and manpower. The number of construction contracts increase and the complexity of the overall project increases. Most API manufacturing companies (the project owners) do not have the in-house resources to effectively administer a fast track project. The owners typically use resources from a construction management or engineering company to monitor field activities, procurement, cost, and schedule.

XIV. PROCUREMENT STRATEGY

There are various ways of contracting for construction and design management. An EPC or design build contract is the joining of the *engineering* design, *procurement*, and *construction* management functions with one supplier. Owners can elect to hire firms with the in-house expertise to manage all three functions or a design firm and a construction management firm who have joined forces to provide these services under one contract umbrella. The advantage of the single contract is the single point of responsibility between the owner and the supplier.

Some owners will elect to have separate contracts with the design firm and with the construction management firm, with the construction firm responsible for the majority of the procurement. The advantage of separate contracts is that the owner retains control over the design, and the design work can start before the construction management firm is selected.

When selecting a firm to manage the design and/or construction of the project, there are a number of strategy and planning tools that these firms must use to effectively manage the overall process. During the various phases of the project, cost estimates and project schedules are prepared. These estimates and schedules are constantly refined as more details of the project are developed.

At the conceptual stage of a project, the cost estimate(s) developed are utilized for the project's overall business strategy. The initial return on investment (ROI) calculations are based on these preliminary estimates. These estimates and schedules although based on limited data, usually not more than 10% of the total engineering, are critical to achieving the project's overall goals. The selection of the key firms to supply the design, construction management, and estimating services is a critical early project activity. The preparation of the project estimates is usually a collaboration between the outsourced vendors and contractors who are familiar with the construction and design market, and the API manufacturing (owner) who is more familiar with the chemistry or the process and the cost structure of the company.

As the design progresses the estimate is refined. The assumptions made during the conceptual estimates must be evaluated for changes. In the initial conceptual estimate, there were constructability assumptions used to prepare the estimate, such as labor availability, equipment and material deliveries, and the sequence and methodology of the construction work. At each refinement, the level of uncertainty is reduced and therefore the level of contingency for unknowns is also reduced. The overall contingency required for a project is in direct relationship to the level of uncertainty or predictability of the final cost of the project. In the initial stages of the project (10%), a contingency of 25+% is common. At the completion of the detailed design and with the process completed, the contingency should not be required at greater than 10%. In a complex process, an additional contingency is established for the final start-up and validation activities. The level of risk, which drives this contingency, is based on the complexity of the process.

When a project is *"schedule driven,"* it is imperative that a schedule be established early in the project and be utilized during the project. During the project planning, the schedule will grow in complexity and task breakdown as the overall project is developed. The schedule must be a working plan throughout the project. This plan will be updated as new data are known and to reflect the current approach to the overall project execution. An effective schedule must have relationships between the work items. The size and number of schedule activities on a project vary from project to project. Many computer programs exist that can arrange the schedule data in easy to analysis formats. A project critical path method (CPM) schedule, to be effective, must have the necessary detail to show a clear critical path. The bar chart is the most common display of a project schedule (Fig. 8).

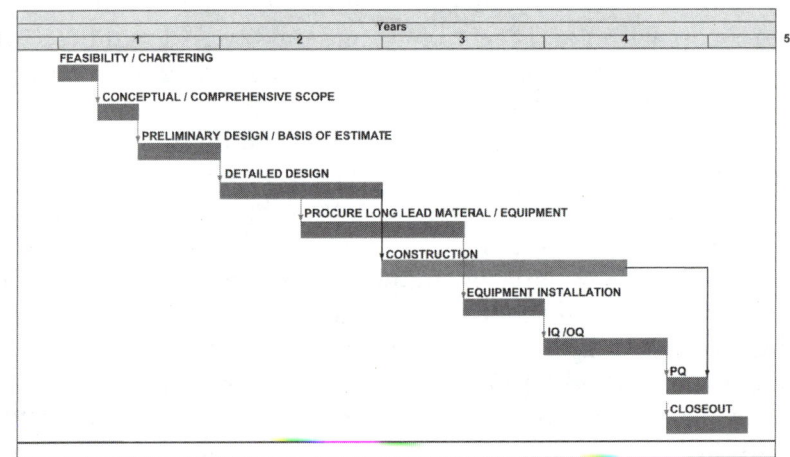

Figure 8 A typical bar chart.

A schedule can also be produced in an arrow diagram, which will graphically show all the activity prior to or dependent on an activity and subsequent activities, those, which follow an activity. This presentation can be very useful in the analysis of how a project can be executed. To monitor a project schedule effectively, the level of activity should be detailed to show the items that should be accomplished during a specific period of time to maintain the overall project completion schedule. The capabilities of the design and construction management firms or EPC firm to produce and monitor an effective project schedule in the complexity required to manage these projects is an important element in the selection process of those vendors.

The selection of the architectural/engineering (A/E) and construction/management (C/M) firms and the contracting strategy with those firms is a function of the schedule drivers. Activities must be worked on concurrently to support the schedule. The suppliers need to provide resources for project planning, early long lead procurement, and conceptual estimating. In many cases, the early involvement of these suppliers is contracted on a reimbursable or cost plus basis. As the project becomes better designed and scoped, the contract between the owner and the A/E and C/M suppliers can become a guaranteed maximum price, lump sum or a reimbursable contract with schedule, and cost incentives. In planning the execution strategy, the resources for the start-up and validation must also be identified early in the process. Many engineering firms have the in-house resources to plan and manage the start-up and validation activities. This is also important to decide when selecting the overall procurement strategy for the project.

With an effective, realistic cost estimate and CPM schedule in hand, the manager of a project can make effective decisions regarding the planning and execution strategy. Many times marketing decisions will dictate the project completion date, which could require additional funds to allow an acceleration of the project by either working overtime or adding additional shifts. When evaluating the final schedule for an API project, the time required for the start-up and validation of the facility is critical to the success of the project. These activities usually start at the completion of construction; however, their duration and requirements make it necessary to start these activities when phases of the construction are complete. Overlapping of these activities will also reduce the overall project schedule. The early planning and strategic development of an overall project strategy will identify schedule opportunities.

The sequence of the construction can then be planned to support the validation. The development of the validation strategy should be developed as part of the overall project execution planning. The strategy for contracting for the validation services is a critical early activity. It is important to identify the process systems that affect or come in contact with the product. These systems must be validated. In a process if chilled water or steam is used to heat the jacket of a vessel but that steam or chilled water never comes in contact with the final product, the utilities will not usually require validation. However, the instruments that control the steam to the vessel jacket will usually require validation. If the controls do not function properly then the product can be overheated or cooled. All pipe systems that transport product must be totally validated.

Another element critical to the completion of a project is the time required for testing of equipment and control systems. In the project validation and procurement planning, all equipment, and systems, which require factory acceptance testing prior to shipment, should be identified. The specifications, and procurement documents should provide for the required testing and identify documentation of the testing procedure for the equipment and control systems prior to shipment to the site. In many cases, with skid-mounted equipment that often has microprocessor controllers, a significant amount of the IQ documentation and verification can be accomplished at the factory. This preplanning will save the overall project schedule. The testing, documentation, and validation activities are then accomplished in parallel with construction activities. Equipment problems are flushed out at the factory prior to the installation in the field. This pretesting at the factory can increase the productivity of the construction installation and reduce the overall project schedule.

In the development of the process automation and control system, the required testing of that control system and the factory-assembled components, and the process simulation program must be established with the general functional specifications. In an API facility, many of the control systems perform process functions that require strict validation. The functional description for the automation system should require a complete factory acceptance test (FAT). This test should simulate the entire process and process failures and alarms. The FAT should also check and verify that the control system cabinets and controllers operate as designed. The factory acceptance testing of the process automation system prior to shipment and installation in the field is a critical step in the validation and start-up of the facility.

XV. CONSTRUCTION MANAGEMENT

As the project transits from the design stages to the construction phase of the project, the construction management plan and subcontracting strategy are developed and finalized. The subcontracting strategy has been completed early in the overall construction plan. Design documents must be prepared to support the schedule of the construction activities and the submittal of vendor drawings and documents is necessary to complete the design effort. Consistent with equipment, early contracting of the process automation system will start the submission of vendor drawings to the engineer for both approval and for inclusion into the electrical and instrumentation drawings.

There are many components to the facility, which may not have a long delivery but may be critical to the completion of the various design and construction packages. Control panels, IO cabinets (the termination cabinets for the input and output wiring of instrumentation) and uninterrupted power supply (UPS) systems need accessibility, which can affect the architectural layout of support space and control rooms. Instrument details are needed from the vendors to finalize the electrical drawings and pipe fabrication drawings. Many times these smaller components will be included in the contract to be purchased by the subcontractors, rather than the API manufacturing companies (the project owners). The construction management firm and the design firm must coordinate these details to insure that the information necessary to complete the design is passed from the vendors to the designers to support the overall project schedule.

The development of a detailed construction schedule is necessary for the coordination of the various contractors and monitoring the progress of the construction. A resource-loaded schedule for construction activities is necessary. Labor requirements must be evaluated from both availability and the density within the construction project. A general rule of thumb is to plan on one construction worker for every 200 gross square feet of building. In the case of compact process facilities, this can require spot density of one worker every $150\,ft^2$. In the electrical and instrumentation trades, this may be as tight as one worker every $100\,ft^2$ of gross floor space. The ability to fabricate pipe skids and systems, off site must be considered both to control labor density and to support the overall schedule goals. The construction schedule must be detailed enough to be able to show the different critical paths but simple enough to be understandable by the various trade workers.

The construction manager's involvement during design development is critical for constructability reviews to be completed at all phases of the design. There is a constant review of how the facility will be built and whether or not the design is practicable. Typical questions include: in what sequence will the facility be constructed? Does the design allow the different trades to complete their work without blocking other trades? Any contractor, who must remobilize to complete their work, will add schedule to the project and cost.

Final documentation for a validated manufacturing facility is critical to the success of the project. Documentation and certification of the work is the responsibility of the construction management team. The format and items required for certification should be included in both the design documents and the validation plan. Almost all documentation provided by vendors is available in electronic format. The computer design programs used to provide this documentation should be established in the early stages of the project and validation plan. The construction manager insures that the subcontractors and suppliers comply with these requirements. Weld documentation and instrument calibration are two of the most common certifications required of the contractors.

XVI. START–UP ACCEPTANCE

The start-up and operational acceptance of any process facility is a complex undertaking, which requires early planning. As systems are started and functioning, they can be reviewed with the operation personnel. Training of the operations personnel for a process facility should start during the final stages of the construction. Normally, training and field start-up support is included in the purchase contracts for all major equipment and system suppliers. The training consists of classroom theory of the equipment or system, the required maintenance, and their intended functions. Classroom training should occur prior to completion of installation of equipment. The operations personnel will then be familiar with the equipment and have participated in validation of the systems.

In an API facility, the construction manager is tasked with coordinating the overall construction and installation of the equipment within the process facility. The installation of the interconnecting components must be installed in a quality manner. Rework takes time and delays the entire process.

XVII. PROJECT TURNOVER AND INSTALLATION QUALIFICATION

The installation qualification (IQ) of a facility is the verification that all the components within a facility that have "direct product impact" are installed correctly and in accordance with the design specifications. There must be supporting documentation that all components have been installed and that all instruments have been properly calibrated. The validation plan defines the methodology for preparing and executing the IQ. It also provides guidance with respect to IQ acceptance criteria, the use of support documentation, and responsibilities. The construction and validation team should perform a walk down of the completed system. The walk down is a review of the completed installation of the construction against the design documents. Usually, a redlined approved for construction (AFC) P&ID will be generated and used as documentation of the completeness of the installation. The walk down should also yield a final punch list of incomplete items.

An example of typical items that would be verified in the IQ phase of validation and in the commissioning of the remaining facility:

- verify that the installation of the equipment complies with the design specifications;
- verify that all required equipment, piping, electrical and instruments are installed;
- redlined AFC P&IDs are completed;
- vendor documentation is available for all equipment;
- full loop checks are completed;
- all necessary utilities are connected and ready for operation.

Any changes from the original P&IDs that are noted during the redlining function of IQ must be documented through a change control procedure, which is detailed in the validation plan. These changes, if approved and accepted, must then be incorporated into the final "as-builts" of the facility. It is critical that the start-up team, maintenance, and the operations have complete accurate documentation of the final as-built configuration of the facility and the process. At the completion of the construction, the construction team for maintenance and operations, and the start-up team must assemble a "turnover package". This package will have at a minimum:

- as-builts drawings of the entire facility;
- preventive maintenance information and requirements on all equipment;
- suggested spare parts for all equipment, including delivery time and pricing;
- vendor manuals on all installed equipment.

The start-up of an API facility is a complex operation that requires early planning and a complete understanding of all the components of the facility. An operations manual with spare part and regular maintenance procedures must be in place on site for every component in the facility. A list of spare parts required for start up should also be assembled and these materials stored on site for use as required during the start up of the facility. The most critical step in developing a complete start-up plan is to develop standard operating procedures for the facility (SOPs). A complex API process facility will require SOPs be developed on every aspect of the operation and including the maintenance procedures for the facility. The SOPs are ultimately the responsibility of the operations team for the facility, but may be developed by an engineering firm or consultant who is initially more familiar with the specific details of the facility and the equipment. An important SOP at this phase of the project is the procedure for the start-up and methodology for shut down procedures of the process. These procedures must show the order in which the equipment should be started and stopped, the setting of valve dampers, instruments and controllers to avoid damage to any equipment. For the initial start-up, which is normally done with "water batching," it may be necessary to compensate for the difference in the specific gravity of water to the process when the facility is in full operation.

XVIII. CONCLUSIONS

- The chapter outlines the steps required to design and construct a new API facility
- The business case
- Process development
- Design
- Execution strategy and planning
- Procurement strategies
- Construction management
- validation, start-up, and project turnover

In the review of the API facilities, distinct differences between the requirements and those of a fine chemical facility exist. The product is manufactured under cGMP guidelines. There are validation requirements for the facility, which document how the facility was constructed, how the facility and process will be operated, and how the facility will be maintained. In most new product introductions, the primary driver for API projects is the speed to market. The products have critical marketing considerations and schedule is of paramount importance, consequently at the risk of higher cost to construct. Because of the differences in the requirements for API facilities vs. fine chemical facilities, both in the construction and final operation, the initial planning strategy for the project is critical to the success of the project.

The design and construction professionals must be knowledgeable in the specific cGMP requirements for API facilities. The construction personnel must plan for the proper documentation of the facility throughout the construction process so that the final facility can be validated and ultimately certified to manufacture product. The design and construction of an API manufacturing facility is a large investment of time, resources, and capital. The proper planning up front and the diligent effort to evaluate the economic options and interface these options with an overall project schedule will produce a facility that operates as intended and returns the predicted profit on the investment.

REFERENCES

1. ICH Q7A, International Conference on Harmonization. Good Manufacturing Guide for Active Pharmaceutical Ingredients. Recommended for Adoption at Step 4 of the ICH Process on November 10, 2000.

2. United States Pharmacopoeia (USP) 23 General Information. Water for Pharmaceutical Purposes. Chapter 1231. Rockville, MD: US Pharmacopoeia Convention, 1995.

3. 21 Code of Federal Regulations (CFR). Current Good Manufacturing Practices for Finished Pharmaceuticals. Part 211.

5

Regulatory Affairs

JOHN CURRAN

Merck & Co., Inc., Whitehouse Station, New Jersey, U.S.A.

I. INTRODUCTION

As we've entered the 21st century, regulatory agencies in the United States (Food and Drug Administration, FDA), Europe, and around the world are placing an increased emphasis on the manufacture and control of the active pharmaceutical ingredient (API), or as it has historically been referred to, the bulk drug substance. With the growing requirements for chemistry, manufacturing, and control (CMC) documentation to support an original marketing application, as well as the advent of agency preapproval field inspections, virtually all aspects of the development and scale-up of the API are subject to regulatory review. Additionally, stringent regulations exist, covering changes to the manufacture and control of the API following approval of the marketing applications, and field inspections are frequently carried out by agencies to ensure ongoing compliance to the marketing application and to current good manufacturing practices (cGMPs).

Given the extent to which the current regulatory requirements govern development, registration and maintenance of CMC information for the API, the preparation of accurate and complete documentation has become more critical. It is increasingly necessary for the regulatory area responsible for CMC within a company to develop the expertise needed to successfully register and maintain appropriate information and documentation on the API, and to assure that changing regulations are tracked, understood, and properly implemented. This chapter is intended as an overview of current API regulations as they exist in early 2004, and is designed to guide and assist the CMC scientist in developing such expertise. While it is not the intent to focus only on the regulations published by the U.S. FDA, it is clear that in most cases, FDA requirements for API are the strictest and most comprehensive. Hence, satisfying FDA requirements often ensures that sufficient information and data exist to satisfy any worldwide regulatory agency.

This chapter presents an overall summary as a "snapshot in time" for an ever-evolving arena in the pharmaceutical industry, and the information contained herein is meant to supplement, not replace, the many excellent guidance documents published and maintained by regulatory authorities, worldwide, as well as the comprehensive documents published by the International Conference on Harmonization (ICH).

The primary focus of this chapter will be on conventional, low-molecular-weight APIs manufactured by chemical synthesis. It is recognized that APIs also are derived from fermentation, proteins, peptides, etc. Specific regulations and guidelines exist for these compounds, which will not be covered in this work. This chapter will also focus mainly on the regulations as they apply to innovator companies seeking approval for new chemical entities. Separate but similar requirements exist for the development and maintenance of drug master files (DMFs) submitted by bulk chemical manufacturers that supply APIs to the industry (note—DMFs for APIs are generally accepted in the United States, Canada, and the EU; other countries have varying requirements regarding DMFs).

Please be aware the existing API regulations and guidelines published by the U.S. FDA and other regulatory agencies are not comprehensive, and are often subject to interpretation by the company as well as the regulatory reviewer. It is therefore important to focus on sound scientific reasoning, supported by analytically valid data, in the preparation of original and supplemental regulatory filings. The ability to clearly communicate the science and supporting data can be a significant challenge for the CMC specialist.

II. REQUIREMENTS FOR SUBMISSION OF REGULATORY CMC DOCUMENTS

A. Investigational Compounds

An increasing number of worldwide regulatory authorities require the submission of relevant information supporting proposed clinical trials prior to the introduction of an experimental compound into man in their country. The terminology used for these investigational submissions differs from agency to agency. In the United States, the documentation is called an Investigational New Drug Application (IND); in the U.K. its a Clinical Studies Exemption (CTX); in other European and international markets, the term used is Clinical Studies Authorization (CSA); and in Canada the document is referred to as an Investigational New Drug Submission (INDS). Generally, the document covers a specific clinical program for a desired therapeutic indication in a target patient population, and must be kept current throughout the clinical development program. Modifications of the indication or target population often require a separate original investigational application.

The original investigational applications and subsequent updates are formally reviewed by the agencies. For original applications, the clinical studies may typically be initiated after a prescribed time following submission, unless the company is informed otherwise by a particular agency (e.g., clinical hold). The agency reviewers may and often do submit questions to the company based on the investigational application, and should clearly communicate whether studies may proceed prior to the resolution of the issues. Frequently, guidance for the ongoing development program is provided by the agency through these questions. Formal responses on all issues should be provided to the agency in an expeditious manner. Often, the responses include commitments for additional investigations as development progresses.

This investigational documentation typically includes CMC information on the chemical characteristics, manufacture, control and stability of the API, and any formulations planned for evaluation in the clinic. For early development candidates, often a brief overview of the API synthesis and summaries of the characterization of the compound and applicable specifications (tests and acceptance limits) are sufficient to allow initiation of clinical trials. As the development program progresses and the compound is to be introduced into larger numbers of patients, more detailed supporting documentation is generally required for the API. Significant changes in the manufacture, characteristics, or controls for the API must be communicated to the regulatory agencies prior to use of the material in ongoing clinical studies, through updates or amendments to the investigational application. Periodic updates documenting other, less critical changes, should also be submitted during the clinical program (e.g., on an annual basis). Often, the updates to the investigational filings provide useful references for generating historical background information on the development program for inclusion in original marketing applications.

B. Application for Marketing

Research-oriented companies will evaluate a number of investigational compounds for potential therapeutic indications. A majority of the potential candidates do not survive the safety and efficacy studies conducted as part of the clinical development program. Those that are found to be safe and effective toward a specific indication must be registered with regulatory agencies worldwide, prior to their being made available for sale in that market. The process by which compounds are submitted for approval to market is very similar throughout most of the world. Specifically, a detailed application must be submitted to the recognized regulatory authority in the country, and that authority reviews the contents and renders their acceptance, conditional acceptance, or rejection of the application. As is the case for the investigational application, several different names are used to describe the marketing application in various countries. These include an NDA in the United States, an NDS in Canada, and a Pharmaceutical Dossier in Europe.

While the registration procedure is similar and the recent adoption of the common technical document (CTD) has begun to standardize the format for regulatory submissions, the content of the quality (CMC) section of the marketing application required by different countries, particularly as it relates to information on the API, varies significantly. In a number of countries, very limited, if any, information is required on the manufacture and control of the API, while in others (e.g., the United States, the EU, Canada, Israel, and Australia) very detailed information and supporting data are required on the characterization, manufacture, control, and stability of the API. A subsequent portion of this chapter covers the information supporting the API to be included in marketing applications for these concerned markets.

Following submission of the application, certain agencies will perform a high-level review of its content to assure that all basic elements are contained in the submission. Once the application is considered accepted for filing, the reviewing chemist or authority from each agency will perform a very detailed evaluation of the CMC documentation, and where appropriate will provide specific questions or comments on the content of the documentation. The CMC questions often seek clarification or additional information or data on specific items, or state concerns the reviewer has with the content or conclusions provided for certain aspects of the application.

Formal responses to each question must be provided to the agency. Timing for responses varies from country to country, but generally the rapid submission of complete responses is desirable to both the reviewer and the company. In most instances, delays in responding to the questions will result in a delay in the approval timing for the marketing application. It should be noted that knowledge of, and adherence to the expectations documented in published agency guidelines generally could minimize the number and/or severity of chemistry questions received on a specific application.

Once the concerns on all aspects of the application, including CMC, are addressed to the satisfaction of the reviewing authority, an official "action letter" is typically provided to the sponsor of the application, to formally allow marketing of the product in that country. Sometimes, conditions for approval are stated in the letter. These conditions should be specific with respect to their impact on the marketing of the product, and often must be satisfied before product is sold.

In the mid to later part of the 1990s, procedures were established to allow for more timely review and approval of marketing applications in the both European Union and the United States. The review process to be used and the timing for approval are defined by the local regulations, and are dependent upon the immediate therapeutic need for the product. The EU mutual recognition and centralized procedures and the U.S. Prescription Drug User Fee Act (PDUFA) will be discussed in more detail later in this chapter.

C. Postapproval Requirements

Following approval of the marketing applications, it is necessary that the CMC information on file with each regulatory agency remains current and accurate. Unfortunately, there are a wide variety of mechanisms that must be followed in the various countries/regions to communicate changes that are required or desired post approval. The mechanisms to be used are often linked to the nature of the proposed change, and its potential to impact the quality (chemical and physical) or safety of the API and ultimately, the quality, potency, safety, or efficacy of the final drug product. Changes having a moderate to significant chance of impacting any of these characteristics generally require approval by the agency prior to the marketing of product containing API made or tested by the changed route. In certain markets, changes having a minimal chance of impacting these characteristics can generally be implemented (i.e., product using API made/tested by the changed route can be marketed), and the agencies are simultaneously or even subsequently notified of these changes via an appropriately defined mechanism. Further details on evaluating and reporting postapproval changes are provided later in this chapter.

Of critical importance in the maintenance of registered information is the existence and implementation of strong change control procedures. For the API, procedures should be in place to address changes to the manufacturing process, (controls and parameters), specifications (analytical test procedures and acceptance criteria), equipment cleaning procedures, raw materials, and/or their acceptance criteria, packaging, etc. These procedures should be consistent with current cGMPs and are often the focus of agency inspections. Defined change control procedures should also be included as part of supply agreements with certain vendors (e.g., suppliers of key starting raw materials and packaging components), since changes made by these suppliers could result in the need for regulatory submissions by the user, which potentially could require prior agency approval.

III. CONTENTS OF REGULATORY SUBMISSIONS—API SECTIONS

A. Content of Investigational Applications

As previously mentioned, a number of regulatory agencies, worldwide, require information on the characteristics, manufacture, control, and/or stability of any investigational API intended for experimental use in man, prior to initiation of clinical trials in their country. The information required is intended to provide the reviewer with a general background and understanding of the compound to be investigated, and is considerably less detailed than that required as part of applications for formal marketing approval. Regardless of the intended reviewing agency or the stage of development for the compound, the main purpose of the investigational application is to demonstrate that the API to be introduced into man is adequately safe and is properly controlled. The exact information required in an investigational application varies from country to country, but generally consists of all or some of the items below:

- General information on the compound
- Description of key chemical and physical properties and characteristics
- Proof of chemical structure
- List of manufacturers
- Method of manufacture (minimally a process flow diagram)
- Specifications (methods and acceptance criteria) for the finished API
- Discussion of impurities and degradation products
- Analytical test results
- Information on the analytical reference standard
- Description of the container/closure system
- Stability data

The level of detail required in the above sections can vary significantly from country to country, and based on the stage of development for the compound. As the development program progresses, more detailed information is generally included in the investigational applications, particularly for the more sophisticated countries. It should be noted that investigational applications may be submitted to a number of different countries at different stages of the same development program, to support individual clinical trials to be conducted in that country.

For certain countries, there exist published guidance documents from the regulatory authority describing the expected contents of the investigational application. This guidance should generally be followed when preparing an application, with any significant deviation from the guidance described and justified. For some sections of the application, the level of detail for information to be provided is not well specified in the agency regulations or guidelines; rather, experience with the respective reviewing agencies should dictate the detail provided. The typical contents of the sections, which comprise an investigational application, follow:

General information—This section should contain the full chemical name, established in accordance with a recognized nomenclature system (e.g., IUPAC); the molecular formula and molecular weight; the stereo-specific chemical structure of the API; plus any internal codes used to designate this compound within the document.

Description of key chemical and physical properties/ characteristics—A physical description of the compound (color, form, and appearance) should be stated. A discussion should be provided on existence of polymorphic forms, solvates or hydrates of the molecule supported by appropriate data (e.g., thermal analyses or x-ray powder diffraction testing). Available solubility data, specific (optical) rotation values for chiral compounds, the partition (distribution) coefficient, acid/base dissociation constants, pH, and hygroscopicity data are generally also included for the selected form of the API.

Proof of chemical structure—Evidence supporting the structural assignment for the API are provided, typically including appropriate spectroscopic or spectrophotometric evaluations and interpretations. At the investigational stage, direct proof of stereochemical conformation may be provided, but generally is not required.

List of manufacturers—The complete names and addresses are provided for all sites that have or will be involved in the manufacture and testing of API for use during the development program. Typically, the names of suppliers of starting materials (i.e., compounds which impart a significant portion of the structure of the final API) as well as auxiliary raw materials used in the synthesis are not given.

Method of manufacture—The contents of this section will vary greatly depending upon the stage of the development program, and the intended country to receive the application. When a compound is being prepared for early clinical trials, the process steps are generally not yet well defined and are frequently being modified or optimized. Thus, for original or early-stage investigational applications, even in the some of the more sophisticated countries, it is suitable to submit a flow diagram of the chemical synthesis, possibly accompanied by a brief, qualitative, narrative discussion of the process steps. Initially, little detail is required for controls on the starting materials and raw materials used in the synthesis. Only major changes to the route of synthesis should impact the filed information early in the development program, requiring an amended investigational application prior to use of the material in the clinic.

As the development program progresses and the synthesis becomes better defined, additional detail is typically provided for the process description. This may occur about the time in development when the synthesis is being scaled from laboratory to pilot plant scale equipment. At this time, acceptance criteria should be considered for key process raw materials, particularly the starting materials.

While the practices of individual companies differ, the process description provided for the manufacture of API for use in the preparation of drug product formulations for pivotal clinical (e.g., the definitive bioavailability study) or stability trials should be documented to a level of detail approaching that to be included in the eventual marketing application. This will facilitate comparison of the synthesis of these key developmental batches with that to be used at the final manufacturing site supplying API for marketed product. At this time, it may be practical to include in-process acceptance criteria for key reactions or intermediates, and acceptance criteria for all raw materials used in this synthesis.

Specifications for the API—As is the case with the chemical synthesis, the analytical controls used to monitor the identity, quality, and purity of the API also evolve during the development program. Hence, there is less of a regulatory emphasis placed on those controls used in the early stages of development, compared to those used in the release of material for pivotal clinical/stability trials late in the development program, and eventually those used for the release of API for use in marketed product. Early in development, there is usually a very limited database to establish meaningful acceptance criteria. Thus, the original investigational application may contain fairly broad or even tentative acceptance criteria and brief narrative descriptions of the analytical methods that make up the early API specifications for the compound. Specifications are generally included for physical appearance, assay, impurity profile, water, final step solvents, inorganic impurities, and identity. Specific (optical) rotation should be considered for chiral compounds, and a particle size control may be applied for certain APIs based on the intended formulation use and their physical or chemical properties (e.g., those with low aqueous solubility). The initial assay procedure may be nonspecific (e.g., titrimetric), until a suitable reference standard is prepared and qualified. Generally, formal method validation data need not be included in the original application, although certain country requirements and sponsor company practices may differ.

By the time API is being manufactured for pivotal clinical/stability studies, the analytical controls and the chemical synthesis are generally well established and a larger database exists for the compound. A specific assay procedure would have typically been developed, and the impurity profile procedure would have been optimized to separate and quantitate impurities and degradation products expected from the established route of synthesis. For compounds with limited number of chiral centers, a chromatographic chiral purity procedure may be applicable, replacing the specific rotation control. The need for control of physical properties relevant to the drug product formulation should also be established, where appropriate. Acceptance criteria can be modified to reflect both process capabilities and the enhanced knowledge of the safety qualification limits for the investigational API. These better-defined tests and acceptance criteria should be included in the investigational application, through appropriate amendments, where necessary. In addition, validation data for the analytical test methods may also be expected by stage in the program. The validation data should generally follow the guidelines provided in the general chapters of an applicable compendium (e.g., United States Pharmacopoeia, European Pharmacopoeia, or Japanese Pharmacopoeia) or in the published ICH guidelines (Q2A/Q2B).

Discussion of impurities and degradation products—This section should include a discussion of impurities, which have been observed in development batches of the API, and to a lesser extent those, which are potentially formed through the route of synthesis. The focus of the discussion is generally around reaction byproducts, but should also include key reagents and solvents (e.g., metals, final step solvents), as well as known and potential degradation products of the API. The discussion should be based on the experiences from the batches made up to the time of submission, as tested by the current analytical procedures. The qualification level of the impurities as determined through safety or other appropriate studies should be addressed. As with the previous two sections, the understanding of impurities for an investigational compound is expected to evolve during development. Hence, early impurity discussions may be less detailed, and may involve peaks for observed impurities, which are not yet structurally identified.

Analytical test results—This section should include test results for batches of the investigational API manufactured by the route of synthesis and tested against the acceptance criteria contained in the investigational application. It is the choice of the CMC specialist based on company practices whether a comprehensive tabulation of data for all clinical API batches should be provided. In all cases, pertinent information such as batch size, site of manufacture, date of manufacture and clinical use should be provided for each batch.

Information on the analytical reference standard—Once a suitable reference standard for the API is established, appropriate information on its manufacture and results from characterization studies should be included in the investigational application. Often it is possible to refer to the synthetic route in the application for the preparation of the reference standard, and to add any additional purification steps used. This section should also contain information on other required reference materials (e.g., an impurity standard).

Description of the container/closure system—A brief description of the primary and minimally functional secondary materials used in the packaging of the finished developmental API should be provided (e.g., double high-density polyethylene liner inside a fiberboard container/closure). Consideration should also be given to including appropriate acceptance criteria for the primary (i.e., product contact) material. The container/closure described in this section should be consistent with that used for stability testing of the compound, or exceptions should be noted.

Stability data—Data from stability studies conducted on the investigational API should be presented, along with the test methods used for the studies, and the conclusions reached from these studies (including the recommended storage temperature and conditions). Stability studies carried out early in development may be conducted against internal protocols, and may focus on establishing suitable storage temperatures and/or container/closure systems. Typically, formal studies for long-term and accelerated stability are conducted on material made for use in pivotal clinical and/or formal drug product stability trials later in the development program. Where practical, these studies should be conducted in a manner consistent with applicable guidelines published by the International Conference on Harmonization (ICH Topic Q1A). This will allow data from these stability studies to be used to support preparation of the eventual marketing application. (Note: ICH Topic Q1A does not specifically cover stability studies to support clinical trials, but does cover studies to support marketing applications.) Results from stress stability studies on the API (e.g., light, oxidative, acid, base, thermal, solution) should also be reported.

As the program progresses, certain agencies now expect that details of the stability protocol used for testing of the API be included in the investigational application.

B. Content of Marketing Applications

The marketing application is the mechanism by which the regulatory reviewing agencies in most countries will grant permission to sell the subject product in their market. The goals for a well-prepared API section of the application are to convince the reviewer that the compound:

- is well characterized;
- is manufactured at production scale using a rugged and well-controlled synthesis, which consistently provides material of comparable or better quality than that used in the development program;
- is tested using validated analytical procedures which show that each batch meets meaningful, justified quality requirements (acceptance criteria) reflecting the quality of the clinical/safety material and the capabilities of the manufacturing process.

Further, the marketing application demonstrates that the API is adequately stable, when stored under defined conditions, to assure that its quality is maintained at least until it is used in the manufacture of the formulated drug product to be marketed.

Historically, there has been a wide range of information on the API expected by worldwide regulatory agencies for inclusion in marketing applications. With the recent publication of guidance on the common technical document (ICH M4Q), the format and high-level content of API information required for a marketing application has been established. While ICH has been developed by and for the United States, EU, and Japan, other regulatory agencies are also following this guidance for new drug applications. Under the CTD, the application should include the following API sections:

- General information

    Nomenclature
    Structure
    General properties

- Manufacturing information

Sites of manufacture
Description of the manufacturing process and process controls
Controls on starting raw materials
Controls on critical steps and intermediates
Process validation or evaluation
Manufacturing process development

- Characterization of the API

  Elucidation of the chemical structure
  Discussion of impurities

- Specifications for the finished API

  Acceptance criteria
  Test methods
  Analytical validation data
  Batch analysis results
  Justification of the recommended specifications

- Reference standard information
- Container/closure information
- Stability

  Summary and conclusions
  Postapproval stability commitments
  Stability data

While the format and high-level content of the application has been established by ICH, the expectations for the exact content (e.g., the type of information and the level of detail required) in the above sections still differ for different markets. Generally, the contents of a marketing application are an extension of the information in the investigational application, with increased levels of detail in certain sections, and clearly defined "bridges," where appropriate, between the material studied in development and that to be routinely manufactured for market. Thus, properly prepared and maintained investigational applications will serve as a good starting point for the preparation of the marketing application for a compound. For APIs purchased from a contract manufacturer, a significant portion of this information may be provided through reference to a DMF, where the regulations allow (e.g., United States, EU, Canada). The typical content for the API sections of the marketing application (or DMF) are well defined in various guidance published by the respective agencies or by ICH. A brief overview of the type of information provided in an API section of a marketing application follows.

1. General Information

Nomenclature—The recommended international nonproprietary name (INN), U.S.-adopted name (USAN) and other national names are included, if available, along with the full chemical name(s) and any internal code names or numbers for the API. Additionally, trade names for the intended formulation(s) and the CAS number for the API may be provided, if established.

   Structure—This section will generally include a full, stereo-specific chemical structure for the API, along with its molecular formula and molecular weight (relative molecular mass). If the compound is a salt or solvate, the molecular weight of the core molecule should also be provided.

   General properties—The information provided for a specific compound may differ based on the exact molecule, but usually includes the following information: physical appearance, thermal behavior, solubility, chirality and specific (optical) rotation, crystallinity and polymorphism, hygroscopicity, partition coefficients, solution pH, and acid/base dissociation constants. The majority of this information is generated during the early development of the compound, and included in the investigational application as discussed above. Under certain circumstances, it may be necessary to evaluate the properties of large-scale pilot batches or even early production scale batches to demonstrate consistency with the properties reported for the early development batches. Any differences in these properties should be discussed as part of the marketing application, and sufficient evidence should be provided to demonstrate the equivalence of the pilot/production material to material used during safety and clinical trials.

2. Manufacturing Information

Sites of manufacture—All sites involved in the manufacture and testing of the finished API and/or key process intermediates for commercial purposes are listed in the application submitted to the U.S. FDA and certain other markets, along with their complete addresses (street addresses are generally required as opposed to P.O. Boxes). Other countries/regions focus only on the manufacturing site of the finished API, while in others this information is not submitted at all. In the U.S. NDA, the sites listed in the application, including contract manufacturing and testing sites, are potential candidates for agency inspection during the approval process. Thus, the sites should be ready to manufacture or test the specified API or intermediate at the time of filing (e.g., the process equipment and/or supporting documents are in place).

If the entire API and/or key process intermediates are manufactured at contract facilities, the information required in this section differs from country to country. In the United States, the contract manufacturer for an API or intermediate is generally identified in the filing, along with specific reference to their U.S. drug master file. A letter from the contract manufacturer, allowing FDA review of the DMF as part of the sponsor marketing application, is also needed. In the EU at this time, the regulations allow DMF reference only for the finished API.

Manufacturing process description—Typically, this section contains a schematic flow diagram of the synthetic route to be used to manufacture the API for commercial purposes, and a textual description of the processing steps. The level of detail for the synthetic process that is required by agencies differs significantly from country to country, and also in some cases from reviewer to reviewer. Certain countries require no information on the manufacturing process, some will accept only a flow diagram, while others expect sufficient information such that they can understand the manufacturing route and are assured that suitable controls are in place to guarantee consistent quality batch to batch. The expectations also differ based on the type of synthetic process used (e.g., more detail would be expected for fermentation processes than for straightforward coupling reactions). The level of detail typically provided by specific companies in their marketing applications also varies greatly. The factors, which impact the level of detail included in the process description, are many, and are often influenced by the philosophies of the individual company; hence, they will not be discussed in this work.

Controls on starting materials—Suitable analytical controls (tests and acceptance criteria) should be included in the application for all raw materials used in the filed manufacturing process description, including those defined as starting materials. The controls required for specific raw materials are dictated by the role the compounds play in the synthetic process. Starting materials, which contribute directly to the structure of the finished API, generally require controls on identity and purity. Raw materials which impact or drive the quality of the API or an intermediate (e.g., a chiral reducing agent) should have controls appropriate to their specific role in the synthesis, while reagents and most solvents may require only identity testing to assure their suitability for use in the process. If the synthesis utilizes more structurally complex starting materials, additional controls such as impurity limits and chiral purity testing may be appropriate. The acceptance criteria for raw materials should be consistent with the demonstrated capabilities of the process. Evidence (experimental data) should exist that raw materials with the specified quality can be successfully processed forward. This evidence would not typically be included in the original filing, but may be needed to respond to an agency question or may be reviewed during an inspection.

A brief synopsis of the test methods for raw materials is generally sufficient, and validation data have, to date, not been required in the original application. For items accepted based on supplier certificates, it is still necessary in most cases to list either the company's or the supplier's tests and acceptance criteria.

It should be noted that currently there are significant discussions ongoing between the regulatory agencies and industry on the development of a suitable, mutually accepted definition of a starting material. Unfortunately, no such definition appears imminent. Many companies elect to review their starting material designations with the regulatory agencies, particularly FDA, prior to submission of the marketing application.

Controls on critical steps/intermediates—Many regulatory agencies expect that companies will monitor quality throughout the process. Thus, they would expect that in-process controls, including tests and acceptance criteria applied to critical steps (e.g., completeness of reaction) and key (isolated) intermediates, be documented in the application. Practices and expectations defining critical process steps and for inclusion of in-process controls in a regulatory process description vary significantly from company to company and from agency to agency. Additionally, in-process testing may be used to demonstrate the removal of certain early step impurities (e.g., reagents or solvents) during further processing, thereby eliminating the need for direct controls applied to the finished API. Usually, in-process controls should trigger appropriate corrective actions if acceptance criteria are not achieved. This can involve, for example, increasing the reaction time, charging additional raw materials, or reprocessing an intermediate. A suitable corrective action should be given for each in-process control provided in the synthesis. Typically, summaries rather than fully detailed in-process test procedures are included in the application, and validation data are not provided.

Recently, more emphasis is being placed on the use of process analytics technology (PAT) for on-line or in-vessel reaction monitoring. The requirements for documenting and supporting the use of PAT are presently being developed by industry and the regulatory agencies.

Process validation/evaluation—Presentation of information in this section of the CTD is a new requirement, and is being actively defined by various agencies at this time. The requirements may be different depending upon the nature of the API synthesis, and the end use of the API (e.g., formal process validation data may be required for the sterility operations for sterile APIs). Consultation of appropriate guidance documents, when available, is therefore recommended.

Process development history—A brief discussion of the development of the synthetic process may be expected in the marketing applications. This section allows the company to establish the bridge from the synthesis used to manufacture early development safety samples, clinical materials, pivotal clinical/stability batches, and API for commercial product. This bridge is particularly valuable for reviewers in the more sophisticated markets who may not have previously reviewed the processes contained in the investigational applications. Again, the level of detail provided for this section will vary considerably.

3. Characterization

Elucidation of the chemical structure—Data are required to support the structural assignment for the API made by the process described in the application. Typically, data are generated using appropriate spectroscopic or spectrophotometric techniques. Interpretations of the data are also included. Additionally, results from a study (e.g., single crystal x-ray) proving the stereochemical conformation of the molecule should be provided. Recently, certain agencies have also been requesting evidence of the chemical structure of starting materials, process intermediates, and key synthetic impurities.

Discussion of impurities—The contents of this section should be suitable to demonstrate to a reviewer that the company understands the origins of, and has in place adequate controls for, impurities which may be present in the API made by the filed process, with a focus on those impurities actually present in batches during the development program. The discussion should demonstrate that the controls in the application are consistent with levels of impurities qualified in safety studies, and with appropriate ICH guidance (Q3A). The level of detail required to achieve these objectives varies depending upon the complexity of the synthesis. As with the process description and process history discussions, the detail of this section is often driven by internal company philosophies.

Regardless of the level of detail, the discussion should include structurally related impurities derived from the route of synthesis, process solvents, reagents and their byproducts (e.g., inorganic impurities), and potential degradation products. For chiral compounds, potential isomers should also be addressed. Typically, the chemical structure and a brief discussion of the formation or origin of the impurity are given. The fate of the impurity in further processing is also included, along with the levels observed in actual batches. A discussion of the methods used to monitor these impurities may be provided or referenced. For impurities that are not directly controlled through the finished API specifications, justification for omission of a direct control may be provided in this section or in the justification of specifications section.

4. Specifications

Acceptance criteria—A listing of the acceptance criteria to be used for release of the finished API is required. These limits should be developed according to established practices within the individual company, taking into account the demonstrated process capability (batch data) and safety/toxicity data for the compound and its potential impurities. Guidance documents have recently been published, which should be considered during the development of the acceptance criteria for the API. Specifically, the ICH has issued guidance documents Q3A—Impurities Testing Guideline: Impurities in New Drug Substances and Q3C—Impurities: Residual Solvents.

Test methods—The analytical procedures used to test the finished API should be provided in the application. The level of detail is again subject to company philosophy; however, sufficient detail should be included to provide the reviewer with a solid understanding of the method. In certain countries, the test methods may actually be run in an agency or contract laboratory, to confirm results on samples provided with the application. Certain tests can be performed using established compendial methods, with the compendial method referenced in the application. Often it is helpful to attach copies of these compendial methods to the application to facilitate review, particularly in countries, which do not follow the referenced compendia.

Analytical validation data—Complete validation data are required showing the analytical methods to be suitable and rugged for the control of the API. Several guidance are available on analytical method validation, including United States Pharmacopoeia (USP) General Chapter <1225>, Validation of Compendial Methods, and the ICH guidance documents Q2A—Validation of Analytical Methods: Definitions and Terminology and Q2B—Note for Guidance on Validation of Analytical Procedures: Methodology. Following these or similar guidance documents should assure suitable validation data in the regulatory application. For tests which reference established compendial methods, revalidation of the procedure is generally not required; however, it may be necessary to demonstrate that the method can be suitably applied to the subject API.

Batch analysis results—Complete batch analysis data should be included for key batches prepared throughout the development program, including, if practical, material manufactured at the site(s) that will supply material for commercial purposes. These test results should be provided against the specifications in place at the time of the release of the batch for its intended use. Often, it is valuable to include retrospective analyses of key batches using techniques established or modified subsequent to their initial release. This could be particularly beneficial for impurities, assuming the impurity profile method(s) were modified during development. It is also expected that information on the manufacture (date, size, location) and use of each listed batch (clinical, safety, stability, market product, etc.) be included in the application.

Justification of the recommended specifications—It is often helpful to present the rationale used for establishing the acceptance criteria proposed for the API in the marketing application. This will provide the reviewers with an understanding of the thought processes used to establish the controls (e.g., are impurity limits based on safety/toxicity studies, analytical results, or both? was ICH guidance followed, which batches were used to set acceptance criteria? were limits determined statistically? etc.). Also, this section can be utilized to support the decision to omit certain direct controls on the API. Providing this information does not assure acceptance of the proposed specifications by all reviewers; however, it may help to focus reviewer comments related to the recommended acceptance criteria.

5. Reference Standard Information

Method of preparation—The preparation of the primary reference standard for the API should be provided, with a focus on the means of purification. The synthesis used to manufacture the batch can be summarized, or referenced to another section of the application, as appropriate.

Characterization data—Full characterization data should be supplied for the primary reference standard, including the results from analytical testing and spectral characterization. The assigned purity of the reference standard should be clearly designated.

If more than one reference standard lot has been made during the development program, and subsequent lots are characterized against the original reference standard, it may be appropriate to provide information on the manufacture and characterization of the original lot in the application. If the reference standard differs from the API (e.g., a more stable salt form or a solution is used), this should be indicated and the rationale provided. Information should also be provided for all other standard materials cited in the application, including, for example, those used to establish chromatographic system suitability and those used for analysis of starting materials, impurities, or intermediates.

6. Container/Closure Information

A description and statement of the composition of the container/ closure system used to store and/or ship the finished API should be provided. Where a nonroutine system is used, a drawing is often very useful. Although the requirements differ, it may also be appropriate to discuss any controls in place for acceptance, particularly of the primary (product contact) components, and possibly the cleaning of the container closure system.

7. Stability

Please note that ICH Q1A, Q1B, Q1C, Q1D, and Q1F provide excellent and current guidance on stability testing that should be considered in the preparation of this section of the application.

Summary and conclusions—A summary should be provided describing how the stability characteristics of the API were determined during the development program. The discussion should include a review of the stability-indicating test methods used in the studies, the batches tested, the storage conditions, and containers evaluated, and the final recommended storage conditions for production material. The use of bracketing or matrixing should be discussed and justified. A shelf life or retest period should be proposed based on analysis of the available data at the time of submission. Testing generally includes both long-term and accelerated studies, using the conditions described in the ICH guidance document Q1A—Stability Testing Guidelines: Stability Testing of New Drug Substances and Products. Additionally, stress studies exposing the compound to acid, base, high temperature, light, and oxidation, should be reported (these studies may be performed as part of the validation of the impurity profile method, to demonstrate selectivity as well as the stability-indicating nature of the method). Solution stability studies may also be performed and reported. The batches used to establish shelf life or retest dating for the API should have been manufactured minimally at pilot scale using a synthesis equivalent to that to be used for preparation of material for commercial purposes.

Postapproval stability commitments—Agencies expect that companies will have in place a procedure for routine monitoring of the stability characteristics of API production material. Recently, several agencies have required that the postapproval API stability protocol be provided in the marketing application. The nature of the commitment contained in the API stability protocol will vary based on the stability characteristics of the material and the practices of the individual sponsor company.

Stability data—Tabulations containing the actual test results from the studies summarized in the stability section of the application should be provided. These results should support the conclusions stated in the document. Any deviations from the recommended acceptance criteria should be noted and explained. Data from stress studies can often be referenced to the analytical methods validation section of the application, or vice versa.

C. Other Documentation Included in Marketing Applications

The CTD format allows other, regional specific information, to be provided in a separate section of the application. The requirements for this section are often described in detail in the individual agency guidance documents. The CTD format also specifies sections for inclusion of references and attachments, as appropriate. Additionally, the CTD includes a quality overall summary (QOS) of the more detailed information provided in the application. The contents of the QOS related to API are fully described in ICH M4Q, and represent a brief overview of the main API section. The QOS may be used during the review of the application by other reviewing disciplines (e.g., clinical, pharm/tox, etc.). The QOS in essence has eliminated the need for a specific expert opinion report and summary tables in the EU; however, it is expected that the contents of the document submitted to the EU have been reviewed and accepted by an appropriate quality expert, and that this expert is suitably identified in the application.

IV. REGISTRATION SAMPLES

Samples of the API, reference standards, and key impurities will be requested by certain agencies so that they can perform the test methods contained in the marketing application. These samples plus certificates of analysis, and sometimes the analytical columns and reagents required for testing, are often sent to the reviewing agency or their designated testing laboratory shortly after submission of the marketing application. The exact quantities required are driven by the test procedures; however, it is not uncommon for an agency to request excessive quantities, particularly of authentic impurity samples. In the United States, samples of the API are also required for retention as forensic samples.

V. THE REVIEW AND APPROVAL PROCESS

The period during which the regulatory agencies are reviewing the contents of the marketing application can be very dynamic, visible, and sometimes intense. Since the goal of the company is to get the product to market in each country in the fastest period of time possible, there will often be pressure for each individual discipline to reach agreement with their reviewers on the final content of their respective sections of the application. This will often require negotiation, clarification, adding information, or providing more detail to address the concerns of the reviewer. Frequently, compromises must be reached. Occasionally, the company must accept undesired decisions by the reviewer in order to gain approval. For the API section of the application, the critical discussions often will focus on the specifications applied to the finished API. Obviously, it is important that meaningful and justifiable test methods and acceptance criteria are approved, and that the manufacturing process is capable of routinely producing material which meets these controls.

A definite strategy should be established for developing responses to agency reviewer questions. It is important that responses are well thought out, provide an answer to the specific question, are based on sound scientific rationale and data, and do not provide excessive additional information that may prompt further questions or concerns. It may be appropriate to seek clarification of the question from the reviewer, particularly in cases where the questions are translated from another language.

While the path of least resistance during the approval process could be to agree with the reviewer comments and recommendations, this approach frequently does not best serve the interest of the company. It is often well worth the effort to defend the original recommendations or conclusions in the application, or to seek a reasonable compromise position. It may be necessary to establish direct dialogue with the reviewer, if possible, to avoid prolonged discussions on a problematic issue.

Questions from regulatory agency reviewers are valuable learning tools, and often aid the company in the preparation of future applications for a particular market or markets. These learnings should be communicated back to the development areas, as much of the content of the application has its origins in early development.

Finally, it is critical to keep appropriate impacted areas (e.g., manufacturing and testing sites, raw material purchasing areas, etc.) within and outside the company aware of the ongoing discussions and the final outcome of the review and approval process. Impacted areas should be directly involved in the preparation, or at least review, or regulatory responses and commitments.

It is well recognized by industry and regulatory agencies alike that review and approval of the application has the potential to be a long and difficult process. In the United States and the EU, regulations have been enacted which seek to limit the duration of the majority of regulatory reviews, provided the information contained in the application meets minimum standards of completeness and acceptability.

The United States adopted a process known as the Prescription Drug User Fee Act (PDUFA), which defines the expected length of time for review of an original application. In exchange for a fee, the FDA agrees to render a decision on the application (approval, approvable under stated conditions, or not approved) within a period of time dictated by the nature of the new drug product. Currently, decisions on applications serving major therapeutic needs (1P compounds) will be targeted for 6 months. Other applications accepted under PDUFA should be reviewed within a 10-month period. Since the timeline is aggressive from an agency prospective, there is more pressure on the sponsor to provide a complete and accurate original application, to be ready for preapproval inspections (PAI) of facilities contained in the application, and to rapidly achieve resolution of issues raised during the review.

In the EU, marketing applications must now be submitted for approval through one of two processes, the centralized procedure (CP) reserved for applications serving major therapeutic needs, and the mutual recognition procedure (MRP), used for all other applications. Under both procedures, the original application is initially reviewed by one of the EU member states. Under the CP, the initial reviewing country (rapporteur) and a corapporteur are selected by Committee for Proprietary Medicinal Products (CPMP) of the European Agency for the Evaluation of Medicinal Products (EMEA). The rapporteur performs an initial assessment of the application, and this assessment is then reviewed by the remaining EU member states. A scientific opinion is rendered on the application 210 days after submission. Questions on the application should be communicated to the sponsor by day 120 in the process. Once approval is reached, the application is considered approved in all EU markets.

The mutual recognition procedure in the EU involves the applicant selecting a reference member state (RMS) for initial review of the application. This review process is less structured in the EU guidelines, and timing may vary depending on the selected RMS and the complexity of the application. The RMS will generally issue incomplete letters during their review, seeking additional information to support their approval of the application. Once the application is approved by the RMS, the applicant generally will update their documentation to reflect the outcome of the approval process (e.g., the CMC documents may require updating to reflect updated or additional information or controls). At the same time, the RMS prepares assessment reports to be shared with the remaining EU member states. The application is then submitted for mutual recognition to some or all remaining

EU concerned member states, which involves a defined 90-day review period. Timing for comments/questions, company responses, discussions, and final action are well defined and quite aggressive. Under MRP, responses to all issues raised by the EU member states must be submitted within 7 days of receipt of the questions, which occurs approximately 2 months following initiation of the process. The final decision on the application (approval, company withdrawal of the application in some or all member states, or binding arbitration) is then made by day 90.

The exact policies governing the EU centralized and mutual recognition procedures and the U.S. PDUFA law are well defined and published, and will not be reviewed here in further detail.

VI. PREAPPROVAL INSPECTIONS

With the recent advent of regulatory agency site inspections during the approval process for marketing applications, an increased emphasis must be placed on the readiness of the site to manufacture and test the specified API or process intermediate. In particular, compliance to current cGMPs and good laboratory practices (cGLPs) must be assured, and supporting data must be available and in good order to substantiate the information submitted in the application. Inspections may occur at any time following submission of the application, but usually will not be initiated until the application has been judged to be acceptable for review by the agency. The exact timing for the inspection may be negotiated, and its duration will depend on the extent of information to be covered, and the observations made.

Most companies will attempt to assure site readiness through internal audits of company and contract facilities prior to submission of the filing. Sites specified in the application should have manufactured the API or intermediate at the time of filing, or at least be suitably equipped and prepared to do so. All process (e.g., master batch records) and control (e.g., analytical quality standards) documentation should be completed and approved. Appropriate standard operating procedures (SOPs), equipment qualification and cleaning documentation, and employee training records should be in order and available for inspection.

The importance of a successful inspection cannot be overstated, as significant concerns uncovered during a pre-approval inspection can delay and even prevent approval of the application. Deficiencies may also be considered indicative of deeper problems within the company, and may have impact beyond the subject filing.

VII. POSTAPPROVAL CHANGE EVALUATIONS

Interactions between the company and the regulatory agencies do not end with the approval of the marketing application. In fact, most agencies place as much emphasis on postapproval activities as they do on approval of the original filing. The reason for this practice is simple; the agencies are charged with assuring the drugs being marketed in their country are and remain safe and effective. Thus, changes to the information approved in the marketing application must be reviewed with or at least communicate to the respective agencies, often before drug product containing API made or tested under the change can be placed on the market.

The content of a postapproval submission reporting a change to the manufacture or control of the API is dictated by the requirements of the different regulatory agencies, as documented in published guidelines. In the late 1990s and into the early 2000s, United States and EU agencies placed a significant emphasis on better defining the impact of changes related to the API, and in communicating the extent of supporting data that should be provided.

The EU published and later modified their guidelines for submission of variations to marketing applications. These guidelines list the type of changes that can be submitted as Type I (minor) variations, and contain the necessary information and documentation required to support the proposed change. A number of the defined Type I (minor) variations reflected changes impacting the API. Any change not specifically defined as Type I in the EU guidelines, must be filed as a Type II (major) variation. Subsequently, the EU further defined minor changes that can be implemented as soon as documentation is provided to the agency (Type IA) and those, which require a brief waiting period to allow an initial broad review before implementation (Type IB). The Type IA/IB designations technically apply to products registered either by the Mutual Recognition or Centralized Procedures; however certain EU agencies are following these same guidelines for older products registered nationally in their country.

Under the current mutual recognition and centralized procedures, variations must be reviewed with, and approved by all concerned member states. The variation is provided under mutual recognition through the reference member state for the original application, and defined timelines exist for review and approval of the variation, based on whether it is a Type I or II submission. The timelines do not include the time required by the application holder to respond to any questions raised by any of the concerned member states.

For variations submitted to the EU, it is necessary to utilize the same CTD format used for new applications, even if the original filing was prepared in an older, approved format (e.g., Part II). This requirement has met resistance from a portion of the industry, in that it implies the need for conversion of existing registration documents to CTD format. European Union has not mandated such conversion; rather they have strongly recommended that the conversion be performed.

In the United States, two guidance documents have recently been published covering postapproval changes to the API. In November 1999, FDA released their Guidance for Industry—Changes to an Approved NDA or ANDA. This document covered API and drug product changes, focusing on the filing mechanism rather than the required supporting data. The November 1999 guidance utilized the premise of assessing change for the potential adverse impact it may have on the safety, quality, or efficacy of the drug product. Changes with significant potential for adverse effect would require formal agency approval before implementation (prior approval supplement). Changes, which present a moderate degree of potential adverse impact, could be implemented upon submission of the required documentation to FDA, or 30 days after that submission [changes being effected (CBE) or CBE-30]. Those changes with minimum potential negative impact could be implemented and submitted in the annual report, an update required by FDA for all active NDAs, based on the anniversary date of initial U.S. approval. Food and Drug Administration published an update to this guidance document in April 2004.

In February 2001, the FDA issued the Guidance for Industry—BACPAC I: Intermediates in Drug Substance Synthesis. This guidance provides filing mechanisms as well as supporting data recommendations for all type changes which impact API manufacture and control prior to the formation of the agency defined "Final Intermediate" in the process. The BACPAC I guidance reenforces FDA's use of appropriate risk assessment in determining the filing requirements for a change. Under BACPAC I, most changes can be reported either in the annual report or via CBE/CBE-30, since changes made early in the synthetic process, or changes to controls on materials early in the synthesis, are logically less likely to have significant potential for adverse impact on the drug product safety, efficacy, or quality. BACPAC I also introduced the concept of being able to evaluate the impact of the change at an appropriate, well-controlled intermediate, which can eliminate the need to take change material forward to the final API to support the supplement.

The Food and Drug Administration is currently drafting for industry review a follow-up guidance (BACPAC II) covering steps from the defined "final intermediate" through the finished API. In general, it is expected that changes made at this late stage will require more regulatory review. Hence, many changes covered under BACPAC II are expected to result in the need for prior approval supplements.

For all postapproval supplements/variations worldwide, the key for the company is to provide evidence that the change does not impact the quality (e.g., impurity profile, physical properties) of the API. It is often sufficient to demonstrate comparability of pre- and postchange material, provided adequate analytical procedures exist to do so. Where this comparability is not achieved, it is necessary to show conclusively that the change does not impact drug product safety, quality, or efficacy, based on either performance tests on the resulting drug product or even by repeat of trials to show bioequivalence to prechange material.

One key factor that dictates often timing for filing of changes to the API synthesis in a number of markets is the need to provide comparability data on material manufactured at full scale, or at least at pilot scale. Since drug product made with the process change material often cannot be released for sale to a market until approval of the change is obtained, the preparation of full scale batches coupled with prolonged approval processes tend to have negative business implications for a company. Thus, changes requiring prior agency approval are generally undertaken only for need or significant long-term benefits (e.g., financial savings).

A tool recently introduced in the EU to facilitate more efficient postapproval change control is the certificate of suitability (CoS). Obtaining a CoS involves the review and approval by the European Pharmacopoeial Commission of the current API characterization, manufacturing and control data for an already registered API. The commission then grants the CoS, which can be referenced in existing marketing applications, replacing the API section of that application. The CoS must be maintained, and therefore changes made to the API must be filed to the CoS. At this time, the EU only recommends the use of a CoS for established APIs; however, legislation has been proposed that would mandate this approach for APIs that have monographs in the European Pharmacopoeia.

VIII. THE FUTURE

It is hopefully evident from the above discussions that regulations governing the API are continuing to evolve. At this time, the requirements still differ from country to country, and these differences have and continue to place an undue burden on the industry. The International Conference on Harmonization has taken an important initiative in developing and publishing meaningful guidance documents for industry, and the positive impact of their activities is referenced throughout this chapter. The published ICH guidelines are generally well accepted by regulators and industry alike, even outside the three primary members (United States, EU, and Japan). There are cases where agencies will enforce requirements beyond those contained in the ICH guidelines; however, the guidelines are now often cited and accepted in a majority of regulatory filings, and are utilized in development of new medicinal products. The ultimate impact of ICH will be judged in the coming years; however, it is already clear that this joint cooperation between agencies and industry is a step forward. We look forward to increased levels of industry/ regulator cooperation in this area in the years to come.

IX. HELPFUL REFERENCES

The Worldwide Web is an invaluable tool allowing today's scientists to keep up to date on current regulatory guidance documents, along with as draft proposals, which govern the preparation of investigational and marketing applications as well as postapproval changes. An excellent FDA website (http://www.fda.gov/cder/guidance/index.htm) contains a comprehensive compilation of the agency guidance documents, arranged by disciplines. This user-friendly compilation also includes the latest ICH guidelines. A similar link (http://www.emea.eu.int/index/indexh1.htm) exists to current EU guidance documents. While guidances and regulations in other countries are sometimes more difficult to find, use of web search tools definitely improve the chances for obtaining the latest information from regulatory agencies, worldwide.

A second set of valuable resources are the official compendia, including the United States Pharmacopoeia/ National

Formulary (USP/NF), European Pharmacopoeia (Ph Eur or EP), Japanese Pharmacopoeia (JP), and other local compendia, of which the British Pharmacopoeia (BP) is generally most comprehensive. These compendia provide insight into the expectations for control of APIs expected by their publishing country, and also generally include general chapters guiding testing, validation, interpretation of results, etc. Most of the compendia are now available via the web, but generally require a license fee per user.

6

Validation

JAMES AGALLOCO

Agalloco & Associates, Inc., Belle Meade, New Jersey, U.S.A.

PHIL DESANTIS

Schering-Plough Corp., Kenilworth, New Jersey, U.S.A.

I. HISTORY

Validation was initially introduced in the 1970s to the pharmaceutical industry as a means for more firmly establishing the sterility of drug products where normal analytical methods are wholly inadequate for that purpose. In following years, its application was extended to numerous other aspects of pharmaceutical operations: water systems, environmental control, tablet,and capsule formulations, analytical methods, and computerized systems. Individuals working with bulk pharmaceutical chemicals (BPCs) were particularly reluctant to embrace validation as a necessary practice in their operations. Industry apologists explained this lack of enthusiasm in terms of differences in facilities, equipment, technology, hygienic requirements, cleaning methodologies, operational practice, and numerous other aspects of disparity that seemingly justified the recalcitrance of this segment of the industry. This view was widespread in the BPC industry through the end of the1980s.

The extension of the concepts that have made validation such an integral part of practices across the healthcare industry to the production of BPCs seems obvious in retrospect. Yet, for many years there existed a general reluctance to introduce validation into BPC activities. While there were some modest efforts, it was not until some time after the biotechnology industry became technically and commercially viable that any significant effort was initiated. The production of biotech products for registration in the United States requires the approval of FDA's Center for Biological Evaluation and Research (CBER). Center for Biological Evaluation and Research required extensive validation of fermentation, isolation and purification processes utilized in the preparation of biologicals (1). An objective comparison of BPC operations relative to those performed in the early stages of biologicals would reveal minimal differences. The production methodologies for many classical BPCs, e.g., penicillins, cephalosporins, and tetracyclines are nearly indistinguishable from those utilized to prepare tPA, EPO, and other biologicals. With this realization, the advent of validation for BPCs was apparent to all, and was increasingly imposed upon the industry.

In 1990, the U.S. Pharmaceutical Manufacturers Association (now called PhRMA) formed a committee to define BPC validation concepts (2). This committee's efforts culminated in 1995 when they issued their finished draft. This document served as a guide to the authors in the development of this chapter. Of necessity, considerable clarification and expansion of the material contained has been necessary to complete this effort.

In the late 1990s, a new term started to appear, first in Europe, but soon it spread across the entire industry—active pharmaceutical ingredients or APIs. Those who first used the new term suggested that it was synonymous with bulk pharmaceutical chemicals or BPCs. Since that time, it has become increasingly common in the industry speak only of APIs. A part of the rationale for this initiative has been voiced as a move towards harmonization. The authors of this chapter do not agree with this change in terminology, as there are numerous bulk pharmaceutical chemicals that have no metabolic activity. Many pharmacologically inactive materials are produced within the industry using facilities, equipment, and methodologies identical to that employed for the so-called APIs, yet with the advent of this new catch phrase are to be seemingly ignored. Our use of the term bulk pharmaceutical chemical is deliberate and is intended to embrace both active moieties, and therapeutically inert materials used as excipients, processing aids, and other materials.

The official requirement for validation of BPC processes was formally established in Guidance for Industry, Q7A Good Manufacturing Practice Guidance for Active Pharmaceutical Ingredients (3). This was the result of a multiyear effort by the International Conference on Harmonization (ICH), which resulted in this harmonized guidance document. This guidance document addresses the subject of validation briefly, and employs the same definition FDA has adopted for other processes (see next paragraph). This chapter provides recommendations for validation consistent with the Q7A guidance.

II. DEFINITION OF VALIDATION

There are innumerable definitions of validation that have been written over the nearly 30 years since its appearance in the pharmaceutical industry. Rather than foster new definitions with the context of this chapter, the authors have chosen to draw upon some of the more widely quoted definitions. The FDA defines process validation as: "Process validation is establishing documented evidences, which provides a high degree of assurance that a specific process will consistently produce a product meeting its predetermined specifications and quality characteristics" (4). This definition is referred to in FDA's subsequent guidance specific for BPCs (5).

III. REGULATIONS

Regulations specific to the control BPCs are a relatively new concept; for many years FDA's policy was to apply a limited enforcement of the subpart 211 regulations for finished pharmaceuticals (6). In recent tears, FDA has endeavored to harmonize its approach to BPC regulation with the rest of the world, and has issued a guidance document that draws heavily on subpart 211 (7). This effort followed the issuance of a pharmaceutical inspection convention document that addressed the same subject in a different format (8).

IV. APPLICATION OF VALIDATION

Some discussion of validation approaches utilized for BPCs is essential to following this chapter. The approaches for BPCs are essentially the same as those utilized for other processes and systems. This discussion serves to highlight the nuances of validation as they apply to BPCs.

V. LIFE CYCLE MODEL

Contemporary approaches to the validation of virtually any type of process or system utilize the "life cycle" concept (9). The "life cycle" concept entails consideration of process or system design, development, operation, and maintenance at the onset. Use of the life cycle helps to provide a system that meets regulatory requirements, but is also rapidly placed into service, operates reliably, and easily maintained. While the "life cycle" is best suited to new products, processes, or systems, it certainly has applicability for existing systems as well. Existing systems that have never been previously validated can be reviewed against the same validation criteria that would be imposed for new systems. While these systems are likely to be deficient with regard to current requirements, the "life cycle"model provides a means for upgrading their programs to be on a par with newly developed systems. This is especially important for bulk pharmaceutical chemicals given that the validation of these processes has lagged behind many of the other areas of the industry, where validation has already been instituted. It is perhaps safe to say that the first validation efforts to be utilized for BPCs will likely be retrospective ones, following the existing system path to enter the "life cycle" model.

VI. VALIDATION OF NEW PRODUCTS

The validation of a new BPC entails practices that parallel those utilized for the introduction of a new pharmaceutical formulation. Thus, a large part of the initial validation effort must be linked to the developmental activities that precede commercial-scale operation. The similarity is such that aspects of reaction, and purification methodologies should be as similar as possible given of course the difference in the scale of the equipment utilized in the commercial facilities. Any differences between the BPC process utilized for the formulation batches used to establish clinical efficacy and the commercial material must be closely evaluated and their impact on the BPC products: chemistry, purity profile, stability, crystal morphology, and other key attributes.

The developmental laboratory has the responsibility for determining optimal reaction conditions including time, temperatures, raw material purity, molar ratios, solvent selection, crystallization method, wash volume, drying conditions, etc. Of primary concern is the identification of critical control parameters, that is to say those that impact quality, purity, safety, and efficacy. The concerns to be addressed in any individual BPC validation program are of course unique to that process, the inclusion or exclusion of any single factor as a consideration in BPC validation is an arbitrary one determined by the authors. Chemical reactions are among the more complex processes to be subjected to validation and the number of critical factors in even a single reaction can be quite extensive. The amount of information which must be generated during development to support a validated BPC process is correspondingly extensive. The necessary information can be assembled into a technology transfer document that conveys the collected experience gained during development to those responsible for the commercial production of the BPC. The success of a developmental organization is better assessed by the quality of the information they convey to document their efforts than it is by the sophistication of the chemistry utilized to make the BPC. The technology transfer document is likely to be of central interest to FDA inspectors during the conduct of a preapproval inspection of the facility prior to approval (10).

VII. VALIDATION OF EXISTING PRODUCTS

At the time this is written, validation of bulk pharmaceutical chemicals is a still relatively new concept for the industry to address. As such the vast majority of BPC products have been on the market without any significant validation in place. As a consequence, the first efforts to validate these products will undoubtedly employ retrospective methods. The trending of results derived from in-process and release testing of these products and processes will serve as the basis for these efforts. Given the FDA's general dissatisfaction with retrospective approaches, it is doubtful that these early efforts will remain the only approach utilized. The use of either prospective (in which three batches must be produced before the process can be considered validated and any of material released for sale) or concurrent (in which individual batches are released while continuing to accumulate data towards a three batch validation) approaches are certainly acceptable, a decision to use those approaches while raising less regulatory concerns will also require a longer time to execute and a larger resource commitment.

The establishment of priorities for validation of a large number of BPC processes generally follows economic concerns, with those products that provide the largest contribution to the firm's profitability being the initial focus of activity. Regardless of how the first validation efforts were completed, the adoption of the "life cycle" model for maintaining products in a validated state is becoming increasingly widespread.

VIII. IMPLEMENTATION

The validation of any process or product relies upon several supportive activities. Validation in the absence of these activities has only minimal utility, as it is only through the integration of these other practices that meaningful validation can be accomplished. Several of these activities are defined in CGMP regulations while others are an integral part of a company's organization structure or are closely associated with "validation" itself (11).

Equipment calibration—The process of confirming the accuracy and precision of all measurements, instruments, etc. to ensure that the measured variable is being accurately monitored. Calibration includes demonstration of conformance to applicable national standards such as NIST, DIN, or BS for all key parameters. This is a universal CGMP requirement across the globe.

Equipment qualification—An outgrowth of "validation" that focuses on equipment related aspects. There is *no* requirement for a formal separation of the activity into distinct elements, such as installation and operational qualification. It has become increasingly common in recent years to combine these activities under a single effort. For the sake of those who still separate the activities individual descriptions have been provided:

Installation qualification—Documentation that the equipment was manufactured and installed in accordance with the intended design. This is essentially an audit of the installation against the equipment specifications and facility drawings.

Operational qualification—Confirmation that the equipment performs as intended entails evaluation of performance capabilities. It incorporates measurements of speeds, pressures, and other parameters.

Process development—The development of products and processes, as well as the modification of existing processes, should be conducted to provide documented evidence of the suitability of all critical process parameters and operating ranges. This effort serves as a baseline for all product validation activities. The integration of development into commercial-scale operations became a requirement with the advent of the FDA's preapproval inspection program (10). The importance of well-documented developmental activities to support subsequent commercial-scale production is essential in the validation of BPCs. It is customary for many unit operations (reactions, separations, catalyst reuse, solvent reuse, etc.) to be initially confirmed on a laboratory or pilot scale, prior to their eventual "validation" on a commercial process scale.

Process documentation—An often overlooked activity wherein the results of the development effort are delineated in sufficient detail in process documentation so that the variations in the process as a result of inadequately defined procedures are eliminated. While master batch records have long been a CGMP requirement, their adequacy is essential to the maintenance of a validated state.

Performance qualification (testing)—That portion of the overall "validation" program that deals specifically with the evaluation (validation) of the process. It includes the protocol development, data acquisition, report preparation, and the requisite approvals. In the distant past this activity was considered "process validation," but over the years the industry has come to realize that "validation" encompasses a broader spectrum of activities and continued use of the word "process" is limiting.

Change control—A CGMP requirement that mandates the formal evaluation of the consequences of change to products, processes or equipment. At least two distinct types of change control exist because of the different disciplines that are central to the evaluation of each (12):

Process change control—A system whereby changes to the process are carefully planned, implemented, evaluated, and documented to assure that product quality can be maintained during the change process. This type of change control is the province of the developmental scientist and production personnel.

Equipment change control—A mechanism to monitor change to previously qualified and/or validated equipment to ensure that planned or unplanned repairs and modifications have no adverse impact on the equipment's ability to execute its intended task. This procedure usually entails close coordination with the maintenance and engineering departments.

IX. BULK PHARMACEUTICAL CHEMICAL VALIDATION

The focus of this chapter is bulk pharmaceutical chemical validation. Aside from the history section, the information presented to this point would apply to most any type of process. That commonality with other older validation efforts is deliberate. Bulk pharmaceutical chemical validation is unique, only to the extent that BPCs are unique. The underlying maxims of success for validation (the knowledge and understanding of the scientific basis upon which the equipment or process is based) are universal. Mastery of the overall approach equips one to effectively employ those concepts in a variety of settings. Some knowledge of the key concerns in the production of BPCs is essential to understanding how the validation of their preparation should be carried out.

Unit operations—BPCs are the result of a series of chemical reactions in which materials, called reactants, are brought together under appropriate conditions whereby the reaction takes place and the reaction product is formed. Under even the most ideal circumstances, the desired product must be separated from unreacted raw materials, byproducts, solvents, and processing aids before it can be utilized in further processing. In the analysis of these processes, chemical engineers have found it convenient to divide the overall process into a series of unit operations (some of which are physical in nature, while others "reactions" are chemical in nature). The unit operations approach is beneficial because a complex many-step process can be separated and better understood as a series of simpler activities (unit operations) that are more easily interpreted.

Among the more common unit operations are mixing, heating, drying, absorption, distillation, condensation, extraction, precipitation, crystallization, filtration, and dissolution. There are other less common unit operations, but the more important aspect is the subdivision of a lengthy process into smaller and more readily understood segments. The benefits to be gained from this approach are obvious, once the underlying principles are understood for a specific unit operation, those concepts can be reapplied in other steps or processes where that same unit operation is employed. In the validation of BPC processes, the ability to use standardized methods for each unit operation can make what would otherwise be an impossible task into a manageable one. The unit operation approach is of such utility that it has been applied in pharmaceutical dosage form manufacturing as well, where the same basic procedures are often encountered, i.e., mixing, milling, filtration, etc.

Physical parameters—A concern that has been sometimes neglected in the preparation of BPCs relates to the control of physical parameters of the end product material. Often the focus of BPC development and processing is on chemical purity and yield, as those aspects tend to have the greatest economic significance. There is relative indifference to physical parameters such as size, shape, and density compared to the seemingly more important concerns such as potency, impurity levels, and process yield. The authors have observed numerous situations where this inattention has resulted in processing problems at the dosage form manufacturing stage. In each instance, it was often the case that the physical parameters of the end product had been virtually ignored in deference to concerns over chemical purity. The FDA's preapproval inspection initiative indicated an awareness that these concerns have come to their attention during the course of NDA reviews and inspections (10).

The most extreme circumstances where physical parameters are of critical importance are for those materials where different crystalline forms are possible. The different polymorphs may have decidedly different characteristics with regard to crystal shape, size, and most importantly solution characteristics. Many important pharmaceutical chemicals exist in more than one crystalline form, and the manufacturer must insure that only the desired form is being produced. One of the major concerns voiced by regulators is the potential hazard in using brokered active ingredients (5). The ability to match the purity profile of a BPC is not sufficient if the crystallization is from a different solvent system or at different conditions. An entirely different material may result, with profoundly different pharmacological properties. The absence of detailed information on the isolation process used may cause difficulties should the real source of the material (the broker's supplier) change.

Chemical purity—Central to the preparation of BPCs are issues relative to the purity of the desired material. Until recently, the only concern was whether the material met the minimum potency requirements. A typical requirement would be a minimum potency specification of 98%. Any lot that had an assay higher than 98% would be acceptable. Awareness that the small amount of material that is not the desired molecule could cause adverse reactions led to the establishment of purity profiles for the molecule. Using a purity profile approach mandates that the firm identify the impurities present. Current FDA expectations are that firms should characterize all impurities that comprise more than 0.1% of the material and perform toxicity testing on any impurity that is at a concentration higher than 0.5% (5). The establishment of a purity profile for a molecular entity assures that process changes, which might result in a change in the byproducts and other materials isolated with the desired material, do not impact the safety and efficacy of the final product.

Analytical methods—As with other types of product validation activities, BPC validation cannot proceed without validated analytical methods. The most significant difference in the validation of BPCs is the number of analytical methods that must be addressed. Analytical methods are needed for each stage intermediate, identifying and quantifying the major byproducts at each stage as well as the desired chemical moiety. Clearly, the scope of the analytical method validation for BPCs represents a larger effort than is normally associated with process validation activities. A comprehensive review of analytical method validation can be found elsewhere in this volume.

Facilities—BPC facilities are vastly different from most other types of facilities in the pharmaceutical industry. The equipment is designed for specialized procedures and as such bears little resemblance to those that might be found in a dosage form facility. Most BPC equipment requires a broader range of utilities and a seeming maze of piping to perform properly. Chemical reactions are sometimes performed at temperatures in excess of 120°C or less than 0°C and required specializing heat transfer fluids to maintain those temperatures.

Many reactions utilize solvents as reaction substrates or in the isolation of the materials. These solvents are sometimes introduced via piping systems that supply the various pieces of equipment. Distribution systems for compressed gases used either in the reaction or to inert the equipment are also common. In many older BPC facilities, it is common to see multiple vessels at different elevations arranged around an open bay. In these facilities, several different chemical reactions might be underway in different vessels for different products at the same time. In a dosage form facility, this type of arrangement would be viewed with some skepticism. In BPC production, the reactions and unit operations take place within completely closed equipment, minimizing the potential for cross-contamination. The difference between BPC and dosage form facilities is most evident in warmer climates. In these areas, the BPC facility may be little more than structural support for the equipment and staging areas for material, with no surrounding building. In effect, the equipment is outside, fully exposed to the environment. For certain BPC processes such as solvent recovery and hydrogenation vessels, the equipment is located outside in even northern climates either because of sheer size or safety concerns. These types of arrangements are not typical for the last step in the synthesis. Isolation of the completed BPC is usually performed in rooms specifically designed for that purpose.

Pure rooms—In the preparation of BPCs, it is common for the last step in the process to be completed in an environment far different from that in which the rest of the synthesis is performed. The term "pure room" is used loosely, there are no regulatory requirements for these rooms and the actual terminology varies considerably from firm to firm. Even without regulatory impetus, some firms have gone so far as to classify their pure rooms at Class 100,000 (EU Class D) or better (13,14). After the crystallization of the BPC, it is important to protect the product from airborne particulates, and other foreign matter that might end up in the finished material. For this reason it is common in many companies to perform a filtration of the active material while still in solution. The filtration removes particulates that may have accumulated in the material up to that point. After the filter, the solution is introduced into the crystallizer in the pure room. The room itself is designed to minimize the opportunity for introduction of contaminants into the bulk material and may or may not be a classified environment. The crystallizer is often subjected to extraordinary cleaning before the start of the process to ensure its suitability for the final bulk isolation. Following the crystallization, the BPC is centrifuged, washed, dried, milled, and packaged in the pure room. It should be noted that BPC processes which use pure rooms are not intended to be sterile, the production of sterile BPCs requires a much higher level of control over the environment, equipment, and methodologies and is described more fully later in this chapter.

Qualification of equipment—The qualification of BPC process equipment including reaction vessels, receivers, crystallizers, centrifuges, dryers, filters, distillation columns, solvent distribution systems, etc. is a well-defined activity. While this equipment is somewhat different in design and operating features, than the dosage form equipment that has been the subject of the majority of papers on the subject, the same general principles apply. Reaction vessels, receivers, and crystallizers differ only minimally from formulation and water for injection tanks. Some BPC dryers are identical to those utilized in tablet departments. Solvent distribution systems are piping systems and may resemble WFI distribution systems. Some pieces of equipment such as distillation columns and continuous reactors may not have counterparts in the dosage form side, but an understanding of the objectives of the equipment qualification should make the development of suitable protocols straightforward.

Configuration confirmation—In multipurpose BPC facilities, the fixed equipment installed may be configured differently for different reactions. In these facilities, campaigns of one reaction may be followed by a reaction for a different product after a change in configuration. Putting aside cleaning considerations for a later portion of the chapter, verification of the systems configuration should be performed. In effect, the reaction train must be requalified at the start of each campaign to insure that the proper arrangement of valves, transfer lines, instruments, and other items are established for the process to be introduced. Some firms run a water or solvent batch, which simulates the process to verify that the proper connections are in place and that there are no leaks in the system. Following the trial batch, the system is then readied for use with the solvents that will be utilized in the process.

Environmental control—The usual concerns relative to the environment in which the production activities are performed are not as significant in BPC manufacturing as they are for the preparation of pharmaceutical dosage forms. The introduction of microbial or particulate contaminants at early stages of the process is unlikely to be of significance. BPC reactions utilize high temperatures, extremes of pH, and aggressive solvents that can minimize the impact of any microbial contamination. Filtration is a frequent part of BPC processing in the form of carbon treatments and other unit operations whose intent is to remove unwanted byproducts, reactants, and solvents. In the course of these measures, incidental particulate contamination is also removed. The use of "pure rooms" as outlined earlier serves to minimize contamination at the last step.

Worker safety—The safety of the personnel who work in the facility is always a major concern. Exposure to toxic substances is greatest when the operator is adding materials to or removing materials from the equipment. The use of air extraction equipment, isolation technology, automated handling, and other means for minimizing human contact with toxic materials is nearly universal. The assessment of worker safety should also embrace exposure to vapor phase hazards, and leak testing of process trains should be performed where hazardous gases are present. Validation of the effectiveness of this equipment is not mandatory from a CGMP perspective, but is certainly beneficial.

Process water—The water used in BPC production is usually deionized water through the early process stages. If the product is isolated from a water solution in its last step, then a compendial grade of water, purified water or WFI may be utilized depending upon subsequent steps in dosage manufacture and the final use of the product. Cleaning of equipment can be performed with city water, provided the last rinse of the equipment is with the same water utilized in the process step. The validation of water systems has been well documented in the literature (15,16).

Process gases—Some BPC reactions utilize gases as reactants, or are performed under a gas blanket. The system may start at either a large high-pressure bulk storage tank or from a bank of gas cylinders. Attention should be paid during the installation of the system to assure that the materials of construction utilized in the system are compatible with the gas being handled. Distribution systems for these gases require qualification, but their similarity to gas distribution systems used in dosage form facilities means that the basic approach is well defined in the literature. For safety considerations particular attention should to be paid to proper identification of process gas lines throughout the facility (see following paragraph).

Compressed air—Air which is classified as breathable should receive an intensive qualification effort especially with regard to the verification of "as-built" drawings, confirmation of proper identification, as well as any safety- and purity-related issues. The emphasis given to breathable air is due to the number of unnecessary deaths, which have occurred in the industry as a consequence of misidentified gas lines. Where air is utilized as a reactant in a BPC operation, it should be treated as described previously under process gases. Instrument air requires the least intensive effort, as the adequacy of the installation can be often confirmed indirectly during the calibration and qualification of the process instrumentation. A single compressed air system could serve as the source for more than one of these air systems simultaneously. In this instance, the advice provided for the most critical application is appropriate throughout.

Jacket services—It is common in BPC facilities, especially those which are reconfigured frequently to accommodate the production of different materials, to have each major vessel equipped with identical utilities, such as chilled water, plant steam, compressed air, and coolant. The use of identical utility configurations on the vessels maximizes the flexibility of the facility, reduces the potential for operator error, and simplifies the design of the facility. The control systems for these jacket services on the vessels would also be identical. Under these circumstances, the qualification effort is greatly simplified through the use of identical requirements.

Solvent distribution—Many facilities use one or more solvents repetitively. In these instances, the installation of a dedicated distribution system for the solvent to the various use points can be justifiable. These systems may be lengthy lines from the bulk storage area (tank farm) to the various locations in the facility where the solvent is required. In some cases, a chilled solvent system may be present to provide chilled washes for use in centrifugation. Depending upon the solvent, specialized piping or gasket materials may be necessary to avoid leaks or corrosion of the system. Qualification of these distribution systems is easily accomplished.

Solvent recovery and reuse—The reuse of organic solvents in a BPC system is widespread, especially given the increased cost of these materials and the environmental difficulties sometimes associated with their proper disposal. This reuse is achieved through defined procedures for the recovery of the solvents from distillates, extractions, and spent mother liquors. Where recovered solvents are utilized in the production of a BPC, the validation of the recovery process is strongly recommended. The validation of the recovery process would include all steps in the process, and confirm the acceptability of the recovered solvent in the processes it will be utilized in. The validation of the use of recovered solvents could be a part of the development of the process. Repeated recycling of solvents could result in the concentration of trace impurities that could adversely affect reaction chemistry. At the very least, recovered solvents should be subjected to release testing and shown to be comparable to fresh solvent. The complexities associated with the validation and reuse of recovered solvents should not be overlooked.

Multiple crops—In the crystallization of some BPCs, multiple crops are sometimes utilized to maximize the amount of material isolated. Even where the cost of the materials being isolated is not high, the ability to increase the overall yield through the preparation of second, third, or even fourth crops is frequently a routine part of the process. A related technique is to recycle the mother liquors without additional treatment from the crystallization to the beginning of the process. Whether through multiple crops or recycling of the mother liquor, both of these processes result in the concentration and/or retention of impurities. The validation of these practices must be a part of the development effort for the process, and reconfirmed on the commercial scale.

Catalyst reuse—Precious and semiprecious metals and other materials are often utilized as catalysts in the conduct of certain chemical reactions: e.g., hydrogenation. While the quantity of catalyst required in any particular reaction is quite low, the cost of these metals is such that recovery is mandated. As the amount of catalyst required to support the reaction is generally supplied in excess it is frequently possible to return the catalyst to the start of the process step without loss in effective yield. The reuse of the catalyst in this manner must be supported by appropriate development work.

Waste treatment—The nature of the materials, byproducts, and solvents utilized in the preparation of BPCs ultimately results in any number of waste treatment problems. The validation of these treatments is certainly *not* a CGMP required activity. Nevertheless, consideration should be given to those activities to insure their reliability. Such efforts can aid in attaining environmental approval for the facility.

X. IN-PROCESS CONTROLS

Bulk pharmaceutical chemicals resemble other types of products validated in the pharmaceutical industry in that they utilize various in-process controls to support and monitor the process through its execution. Typical controls that might be a part of a BPC process include material specifications.

Material specifications—The controls of reactants, solvents, intermediates, and finished materials employ formal specifications for key parameters. The importance of these controls increases towards the end of the synthesis and any of the controls that follow the BPC step are certainly important enough that the efficacy of limits set for these controls should be a major part of the developmental process. Foremost among the considerations in the latter process steps should be the impurity profile of the key intermediates (see following paragraph). Physical parameters (size, shape, crystalline form, bulk density, static charge, etc.) of the finished BPC are sometimes considered less important than chemical purity. When the BPC is formulated in a solid or semisolid dosage form, these physical parameters may assume far greater significance.

Purity profiles—Within the specification parameters, prominence is often given to the establishment of purity profiles for the key intermediates and finished goods. The FDA mandates the identification of all impurities with a concentration greater than 0.1% and generation of safety and other critical information for impurities at levels of 0.5% or higher (5). The establishment of purity profiles for the final BPCs provides for confirmation of the safety of the active material. It is often beneficial to establish purity profiles for intermediates earlier in the synthesis to prevent the carryover of impurities to the finished BPC. The maintenance of the purity profile mandates that a careful evaluation of process changes and potential alternate suppliers of solvents, raw materials, intermediates, and BPCs be made. The analytical method development and synthetic chemistry skills required to obtain the necessary data on impurities meeting the FDA's criteria are substantial. These efforts are well rewarded in an expanded knowledge of the process chemistry and analysis that can assure the quality of the desired active moiety.

Vendor support to validation—A common practice in BPC production is the subcontracting of certain chemical steps to outside suppliers. As is the case with subcontracted production for dosage forms, the owner of the NDA or DMF maintains responsibility for the validation of the process and must secure the cooperation of the subcontractor in the performance of any supportive qualification/validation activities. Agreement to this arrangement should be a precondition to the awarding of the contract to the supplier.

Supplier quality evaluation and audits—Suppliers of intermediates, reactants, solvents, and other materials should be subjected to the same types of evaluation utilized for other dosage forms. The extent of the assessment should vary with the importance of the material to the process. Precedence would be given to those materials whose purity would have an increased impact on the finished BPC. Where the material being produced by the vendor has direct impact on the BPCs quality, as would be the case for chemical intermediates, a more intensive approach is required. Periodic audits of these key suppliers should be a part of the overall quality assurance program.

Sampling plans—Obtaining samples of finished BPCs or their intermediates presents the same difficulties encountered in the sampling of any similar material. When samples are taken of powder or crystalline materials, questions regarding the uniformity of the material being sampled must be addressed before the results of the sampling can be considered meaningful. Bulk pharmaceutical chemicals that are dried in rotary or fluidized bed dryers may be blended sufficiently as a result of the drying process. However, where tray dryers are utilized, a final blending of the dried material may be required before sampling for release to the next stage of processing. In certain instances, an intermediate or finished material will not be isolated as a dry powder but will be released as a solution in an appropriate solvent. Under these circumstances, concerns regarding the sampling of the material are minimized.

Particle sizing—Milling and micronizing are common activities in the final stages of BPC manufacture. These procedures are utilized where the BPC producer has committed to providing a particular particle size for use in the formulation. Given the importance of particle size in many final dosage forms, where present these processes should be validated. Control of the final particle size for finished BPC should not rely on the milling/micronizing step alone. Control over the crystallization procedure is generally necessary to minimize the variation in the material that is to be sized in the mill. It should not be assumed, that the milling/microni-micronization procedure will be tolerant of a wide range of materials and still provide a consistently sized finished BPC product. The uniformity of materials is sometimes improved by passage through a particle sizing procedure or sifter, but this step alone should not be considered sufficient to achieve a uniform mix of the material prior to sampling (see prior paragraph).

Reprocessing—There is occasional need to reprocess an intermediate or finished BPC in order to alter its crystal size, reduce impurities, or otherwise recover off-specification material. Where these processes are utilized, their inclusion in the validation program is essential. FDA requirements on reprocessing and reworking of materials require the validation of any material reclaimed in this fashion. This is most readily accomplished as a part of the developmental process.

XI. CLEANING VALIDATION

A comprehensive discussion of cleaning validation is beyond the scope of this chapter, the reader should refer to other sources on cleaning validation for details of this activity (17–19). Within the context of this chapter, only those aspects of cleaning validation unique to BPC production will be presented. Additional guidance can be found in FDA's BPC Inspection Guide (5).

Boil-outs—Commonly used to clean BPC equipment, boil-outs entail the introduction of the solvent (it could be water) used in the just completed process, and heating it to reflux. The expectation is that the evaporation/condensation densation will result in the dissolution of any residue on the equipment in the solvent. This will remove it from the internal surfaces that are ordinarily inaccessible for direct cleaning and thus clean them. Boil-outs are also utilized as one of the last steps in preparation of equipment for the start of a process or campaign.

Lot-to-lot cleaning—As the production of BPCs often requires that solvents and materials with substantial toxicity must be employed, cleaning of the equipment after completion of the process has the potential for exposure of the worker to those materials. For this reason, it is common in BPC facilities to include some basic forms of waste treatment and equipment cleaning directly into the process in an effort to minimize worker exposure later on. In addition to these measures, many processes include the reuse of equipment and retention of materials in the equipment without cleaning. A typical instance would be leaving a heel in the centrifuge at the completion of the batch, thereby eliminating cleaning of the centrifuge after each batch. The retention of the heel must be validated as it represents a portion of the first batch, which may now become a part of subsequent batches. In fact, each batch in the entire campaign is potentially mixed with material from every prior batch! In this manner, the amount of cleaning required between batches of the same reaction step would be reduced. In those facilities, where a process train is essentially dedicated to the same reaction step over a long period of time, the equipment and process are specially designed to minimize batch-to-batch cleaning of the equipment. There are of course instances where the presence of even trace quantities of finished material at the start of the reaction may create an undesirable outcome, in those circumstances the equipment must be cleaned after the completion of each batch. Sparkler and other filters used to recapture

catalysts, activated carbon used for decolorization and byproducts may require cleaning after every batch.

Campaigns—The production of a number of batches
of an identical synthesis in the same equipment is common
in the manufacture of BPCs. As mentioned earlier in
relation to the qualification of equipment, production in a
campaign mode may require the partial reconfiguration of
the equipment train to allow for a new campaign. This
may be a reaction leading to the same or a different BPC.
To allow for campaign usage, the extent of cleaning required
will generally be far greater than what is carried out
between batches of the same process step. Cleaning limits
for campaign cleaning are generally tighter than those
applied for batch-to-batch cleaning. It is beneficial in campaign cleaning to follow a defined plan for changeover from
one product to another.

Sampling for residuals—In order to determine whether a
piece of equipment has been appropriately cleaned, sampling
is performed. Here again, the particular nature of the BPC
materials makes for a more difficult situation. In dosage form
manufacturing, relatively few of the materials likely to be
retained on the surface of the equipment poses any substantial risk to the worker. In those dosage form processes where
toxic or potent materials are handled, the design of the equipment with smooth surfaces, rounded corners, sanitary fittings, etc. reduces cleaning difficulty. The same equipment
design principles make sampling of pharmaceutical equipment relatively simple due to provisions for access and inspection. The bulk of BPC equipment is designed to operate under
more aggressive conditions, and cannot always integrate the
design features so commonly found in their pharmaceutical
counterparts. Moreover, worker safety becomes a far greater
concern, as the solvents and materials are not conducive to
direct exposure to the employee. Sampling of BPC equipment
may be restricted to fewer locations, and those locations are
generally not in the most difficult to clean or "worst case"
locations. For this reason, the residual limits for BPCs may
need to be far lower to accommodate the uncertainty of the
sampling that can be performed.

XII. COMPUTERIZED SYSTEMS

The application of computerized systems in the pharmaceutical industry is perhaps greater in BPC processing than in any other. Distributed control systems (DCS) have been utilized for many years in the control and regulation of chemical process plants. Their adaptation to BPC preparation is straightforward. The validation of computerized systems in the pharmaceutical industry has been extensively discussed, with the constant recognition that their extensive usage in BPC production was a given (20,21). Industry and regulatory guidance having always recognized this fact, this chapter could not hope to do justice to the subject which has filled several textbooks on its own. The reader is encouraged to follow the recommended practices of PDA, PhRMA, and GAMP.

XIII. PROCEDURES AND PERSONNEL

Where computerized systems are not utilized for the execution of the chemical synthesis, the chemical operator, following detailed batch records is responsible for the operation of the equipment. The batch records must provide for sufficient detail to insure that the worker can safely and properly perform the desired actions. In certain larger process trains, more than one operator will work simultaneously on the same batch. Provided that their activities are closely integrated, there is little problem with this type of approach. The personnel must be trained in their jobs and records of the training must be retained by the firm.

XIV. VALIDATION OF STERILE BULK PRODUCTION

The preparation of BPCs, which must be sterile upon completion of their synthesis and purification, is a common activity in the pharmaceutical industry and increasingly common in biotech processes. The validation of sterile BPCs represents one of the more difficult activities in the entire spectrum of validation. Not only must the final material meet all of the physical and chemical requirements associated with other BPCs, it must also be free of microorganisms, endotoxin, and particulates. In doing so, all of the considerations for validation of BPCs outlined in this chapter must be addressed, with added concern for sterilization, environmental control, aseptic technique, and other subjects associated with the production of sterile products. The following sections address those issues relating to sterile BPCs that are somewhat different from either the validation of nonsterile BPC production or the validation of other sterile materials.

A. Product Sterilization and Sterility Assurance

The predominant method of sterilization for BPCs is by membrane filtration. This filtration will require validation in accord with regulatory expectations. Adaptations to the common filter validation methodologies may be required for certain solvents and/or antibiotic solutions. Subsequent to the filtration step, the succeeding unit operations must be carried out using facilities, equipment, and methods designed to prevent the ingress of microorganisms. The remainder of this section reviews considerations relative to sterile BPC preparation under these constraints.

B. Closed Systems

Central to understanding much of what is presented below is recognition that BPCs, whether intended to be sterile or not, are primarily produced in closed systems in which the reaction, separation, and purification unit operations take place. A joint PDA/PhRMA task force has defined a "closed system" as:

> "system which is designed to prevent the ingress of micro-organisms. A "closed" system may be more accurately defined by characteristics of its operation than by a description of its physical attributes, as these will vary from system to system (22)."

A "closed" system

- is sterilized in place or sterilized while closed prior to use;
- is pressure and/or vacuum tight to some predefined leak rate;
- can be utilized for its intended purpose without breach to the integrity of the system;
- can be adapted for fluid transfers in and/or out while maintaining asepsis;
- is connectable to other closed systems while maintaining integrity of all closed systems (e.g., rapid transfer port, steamed connection, etc.);
- utilizes sterilizing filters, which are integrity tested and traceable to each product lot.

Closed systems provide for complete separation between the environment in which personnel (uniformly accepted as the primary source of contamination in aseptic environments) are located from that in the materials are processed. Theoretically, if a sterile BPC could be processed in its entirety within closed systems, there would no possibility of microbial contamination. In marked contrast to the "closed system" is the "open system", perhaps best defined by what it is not. Essentially, an "open system" lacks one or more of the features of a "closed" system, thus leaving it vulnerable to the potential ingress of contamination. One substantial issue associated with these definitions is establishing that a system remains "closed" over the length of the production campaign.

Facilities—The production of sterile BPCs requires a composite of design features drawn from both sterile dosage form and bulk pharmaceutical chemical production. Ceiling, walls, and floors are composed of materials that can be subjected to frequent cleaning with disinfectants. Pressure differentials are provided to prevent the ingress of contamination from less clean areas into critical processing areas. In order to perform the reactions and separations necessary to prepare and isolate the BPC, processing equipment not generally associated with aseptic environments must be introduced. Centrifuges and crystallizers must be adapted for use in an aseptic area. The finished facility is most certainly a hybrid, as compromises are inevitable to accommodate the essential requirements. The case can be made that if the production systems are perfectly "closed," then concerns relative to facility design required for asepsis would be lessened. The authors are aware of several sterile bulk production facilities in which only a small portion of the system is actually located in an aseptic environment. Certainly, "open systems" must approach the highest levels of aseptic design in order to be successful in operation.

Environmental classification—The environments in which sterile BPC production is executed can vary with the degree of closure provided by the equipment. "Closed" systems as described earlier have been successfully operated in Class 100,000 (EU Class D) or unclassified environments. Systems that are open are generally located within Class 100 (EU Class A) where product is exposed, and surrounded by Class 10,000 (EU Class B or C).

Utilities—There is very little difference between utility systems for a sterile bulk plant and those found in a typical BPC facility. The only differences might be utilities uncommon in a BPC plant such as water for injection and clean steam. The validation requirements for these systems have been well defined in the literature and need little mention here.

Layout—The layout of a sterile bulk facility will again be a hybrid of those found in a conventional BPC plant and a sterile dosage form facility. There will be nesting of classified environments, with critical activities performed in the areas of highest classification. Pressure differentials are employed between clean areas, and those adjacent, less clean areas. The design features are drawn primarily from the dosage form facility model with adaptations to accommodate the generally larger equipment required for bulk production.

Isolation technology—The use of isolators and closed systems for the production of sterile bulks is strongly recommended. As with any aseptic process, the sterility assurance level associated with a sterile bulk material is closely related to the extent of direct human intervention with the material. Isolators and closed systems minimize the need for personnel contact with critical surfaces and thus minimize the potential for contamination of the sterile materials from human-borne microorganisms. It should also be noted that isolation technology can be useful in the containment of potent compounds as many BPC intermediates and finished materials often are. Isolation technology is a rapidly evolving area and the reader is encouraged to stay abreast of current developments (23,24).

Sterilization in place—Closed systems such as process vessels, dryers, centrifuges, isolators, and other items should be subjected to a validated sterilization procedure, which assures that all internal surfaces have been rendered free of microorganisms. Sterilization-in-place (SIP) procedures reduce the number of aseptic manipulations necessary to ready the equipment for use in the aseptic production processes and as such are considered preferable to aseptic assembly of systems from individually sterilized components (25). The SIP procedure should allow the system to maintain sterility until ready for use without aseptic manipulations. Sterilization-in-place procedures could employ steam, gas, dry heat, radiation, chemical agents, or other validateable sterilization procedure.

A brief overview of some of the various sterilization-in-place methods available and their validation follows:

Steam—Primarily utilized for systems composed of closed vessels, with interconnecting piping. It has some similarity to empty chamber studies in steam sterilizers. Important parameters to confirm are appropriate time–temperature conditions throughout the system. Emphasis is placed on the removal of air and condensate from the system, strict adherence to the defined sequence for the sterilization procedure and inclusion of methods for the protection of the system between sterilization and use.

Gas—Utilized for systems that cannot withstand either the temperatures or pressures employed in steam sterilization. Critical parameters for sterilization are time, temperature, relative humidity, and gas concentration. Gases in widespread use include ethylene oxide, peracetic acid, and hydrogen peroxide. Gas sterilization is most often encountered with isolator systems, freeze dryers, and other systems which have limited ability to hold pressure.

Dry heat—Employed in specialized systems where the presence of high temperature for the process is commonplace, i.e., spray dryers, flash dryers, and similar equipment. Confirmation of time–temperature conditions in the equipment is critical to the validation.

Radiation—Radiation sterilization is most commonly utilized for flexible packaging components that can be sterilized while closed prior to filling. The validation of radiation sterilization relies on confirmation of the delivered dose to all portions of the materials, and confirmation of stability after the treatment.

Chemical—Many of the strong acids, strong bases, chemical solvents and other chemicals utilized in the preparation of sterile BPCs have the ability to reliably destroy microorganisms. These materials because of the extreme pH, or other aspects of their chemical structure can effectively sterilize processing equipment. As their use in the system will generally mandate that the equipment surfaces can be exposed to these materials for extended periods of time, their use as a sterilizing method for the equipment is facilitated. Concentration and duration of contact are the critical parameters that must be confirmed in the validation of these treatments.

Aseptic processing—The validation of the sterile bulk process follows the general approach described earlier for nonsterile bulks. The overall process can be divided into a series of unit operations that can be addressed individually or in groups. This approach can be used equally well for aspects of the chemical reaction, purification, physical processing (i.e., milling, sieving, etc.) or aspects related to sterility assurance. A comprehensive treatment of validation methods for validation of aseptic processing for sterile bulks has been developed by a joint PDA/PhRMA task force (22). This document embraces such aspects of the validation as: the use of closed or open systems for processing; materials to use in the conduct of the simulation; sampling and testing of materials; duration of simulation, simulation size, campaign production, and acceptance criteria to be employed. Producers of sterile bulks are already familiar with the contents of this document, and the interested reader is encouraged to read this guidance in its original context.

XV. CONCLUSION

This chapter has provided an outline of validation considerations relative to the production of bulk pharmaceutical chemicals. This is a subject that has only recently become of interest to the pharmaceutical community. The authors while familiar with both validation and bulk pharmaceutical processing have undoubtedly mentioned any number of issues which may not yet be embodied in validation protocols within operating companies. We have included these issues to insure completeness in the presentation, not to suggest that they be included in every validation effort. As time passes, the industry will gain experience with the validation of BPCs and will perhaps exclude some of these issues, while including other aspects we have not identified. Our intent in this effort has always been to integrate common validation practices with the unique aspects of bulk pharmaceutical manufacturing. By no means do we expect this to be the definitive effort on this complex subject. The reader is encouraged to monitor industry and regulatory developments relative to BPC validation, as substantial changes in CGMP requirements for BPCs appear likely.

REFERENCES

1. Food and Drug Administration, 21 CFR, Part 610.

2. PhRMA Quality Control Bulk Pharmaceuticals Working Group. PhRMA Guidelines for the Production, Packing, Repacking or Holding of Drug Substances. Pharmaceutical Technology, Part 1, December 1995, Part 2, January 1996.

3. Food and Drug Administration. Guidance for Industry, Q7A Good Manufacturing Practice Guidance for Active Pharmaceutical Ingredients. 2001.

4. Food and Drug Administration. Guideline on General Principles on Validation. 1987.

5. Food and Drug Administration. Guide to Inspection of Bulk Pharmaceutical Chemicals. 1994.

6. Food and Drug Administration, 21 CFR, Part 211.

7. Food and Drug Administration. Guidance to Industry: Manufacturing, Processing or Holding Active Pharmaceutical Ingredients. March 1998.

8. Pharmaceutical Inspection Convention. Internationally Harmonized Guide for Active Pharmaceutical Ingredients—Good Manufacturing Practice. September 1997.

9. Agallco J. The validation life cycle. J Parenteral Sci Technol 1993; 47(3).

10. Food and Drug Administration. Guide to Inspections of Oral Solid Dosage Forms Pre/Post Approval Issues for Development and Validation. January 1994.

11. Agalloco J Validation—yesterday, today and tomorrow. Proceedings of Parenteral Drug Association International Symposium. Basel, Switzerland: Parental Drug Association, 1993.

12. Agalloco J. Computer systems validation—staying current: change control. Pharm Technol 1990; 14(1).

12a. Agalloco J. personal communications, 1972–1990.

13. Federal Standard 209E. Airborne Cleanliness Classes in Cleanrooms and Clean Zones, September 1992.

14. EU Guide to Good Manufacturing Practice for Medicinal Products. Annex 1—Manufacture of Sterile Medicinal Products.

15. Meltzer T. Pharmaceutical Water Systems. Tall Oaks Books 1996.

16. Artiss D. AWater Systems Validation. In: Carleton F, Agalloco J, eds. Validation of Aseptic Pharmaceutical Processes. New York: Marcel-Dekker, 1986.

17. Agalloco J. Points to consider in the validation of equipment cleaning procedures. J Parenteral Sci Technol 1992; 46(5).

18. Madsen R, Agalloco J, et al. Points to consider for cleaning validation. PDA J Pharmaceutical Sci Technol 1998; 52(6)(suppl). PDA Technical Report #29.

19. Voss J, et al. Cleaning and cleaning validation: a biotechnology perspective. PDA, 1995.

20. Harris J, et al. Validation concepts for computer systems used in the manufacture of drug products. Proceedings: Concepts and Principles for the Validation of Computer Systems in the Manufacture and Control of Drug Products, Pharmaceutical Manufacturers Association, 1986.

21. Kemper C, et al. Validation of computer-related systems. PDA J Pharmaceutical Sci Technol 1995; 49(1)(suppl). PDA Technical Report #18.

22. Agalloco J, Lazar M, et al. Process simulation testing for sterile bulk pharmaceutical chemicals. PDA J Pharmaceutical Sci Technol 1998; 52(1). PDA Technical Report #28.

23. Akers J, Wagner C. Isolation Technology: Application in the Pharmaceutical and Biotechnology Industries. Interpharm 1995.

24. Coles T. Isolation Technology—a Practical Guide. Interpharm 1998.

25. Agalloco J. Sterilization in place technology and validation. In: Agalloco J, Carelton FJ, eds. Validation of Pharmaceutical Processes: Sterile Products. Chapter. New York: Marcel-Dekker, 1998.

7

Quality Assurance and Control

MICHAEL C. VANDERZWAN

Pharmaceutical Technical, Roche Pharmaceuticals, Basel, Switzerland

I. INTRODUCTION

The quality of active pharmaceutical ingredients (APIs) is defined as meeting the appropriate specifications for the API <u>and</u> being produced in a facility compliant with ICH guidelines "Q7A" and FDA's current good manufacturing practices (cGMPs) regulations. Most countries regulate the manufacture of APIs. These regulations require a total systems approach to assuring an API has the appropriate level of quality. All components in this system must be properly designed, validated, maintained, and operated to allow the manufacturer to assure the API consistently meets quality requirements. The general components of the system are the process, facilities, and the people. This chapter concerns these components, as well as the product quality itself, the regulations, and the quality management (QM) department.

A. The Product

The quality of an API is determined by two factors: its conformance to pre-established specifications and whether it is produced according to a documented validated process in a cGMP compliant facility.

The API must possess appropriate chemical and physical attributes to assure that it delivers the intended pharmacological effect. The chemical attributes describe the appropriate purity and impurity limits. Impurity specifications are established from clinical toxicological studies and are also based on reasonable minimums expected from regulatory authorities and consumers. The physical attributes describe the necessary characteristics for reliable pharmaceutical processing into final dosage forms. These attributes are determined by empirical evidence from formulation trials to produce uniform and stable dosage forms of adequate bioavailability.

B. The Process

The quality of the API is designed into the molecule through the development of the full manufacturing process, from the laboratory scale synthetic process through to end product. The synthetic process must be designed to minimize impurities, especially those that prove difficult to remove in the last step. Thus, through effective process development, yields are maximized, waste is minimized, and impurities are not formed, eliminated, or certainly minimized. The specific controls used by the developmental chemist to produce the high-yield, high-quality product must be documented; this documentation forms the basis for the *proof of concept* and for the *validation report*. In nearly all countries today, regulatory authorities require the API to be produced from a documented process that reliably meets all appropriate specifications. This was strengthened by the issuance and adoption of the International Conference on Harmonization Tripartite Guideline of Q7A "Good Manufacturing Practice Guide for APIs." The European Union, the Japanese Ministry of Health and the United States Food & Drug Administration adopted the guide.

C. The Facilities

The facilities in which APIs are produced are also addressed in this chapter because a component of quality of an API is that it be produced in cGMP-compliant facilities. Those components of the facility governed by cGMP are therefore part of this chapter. The essence of cGMP for facilities or, for that matter, any aspect of API manufacture is that the facility performs as designed to assure the quality of the product. Further, the performance characteristic must be documented, and management must demonstrate the facility continually performs as designed. Performance control monitoring, preventative maintenance, and carefully controlled and approved repairs or changes to facility components are all considered part of assuring quality of APIs.

D. The People

The people who produce the API are considered a critical part of the system and, as such, become part of the requirements for quality of APIs. To do their jobs effectively and to assure quality of the API, they must be properly trained and equipped. Qualified personnel must conduct the training; the equipment must be of proper design and function. The supervisors of people manufacturing APIs must also be properly trained to do their jobs. Finally, there must be an adequate number of people to allow sufficient time to perform these responsibilities in a satisfactory manner.

E. The Quality Management Department

As in most any other manufacturing enterprise, there is a quality control and/or a quality assurance department. Today, these departments are usually combined into a QM department. The role of the QM department has also advanced from "check-test-decide" responsibility to being an equal partner

with manufacturing and engineering to manage and improve
the quality of the entire process and system.

For APIs and drug products, the QM department, through
its quality assurance arm, still has the responsibility vested in
it by regulations to release all products for use and eventually
to the market. As a component of the system to produce APIs,
the activities and responsibilities of the QM department are
also a component of product quality. Most cGMPs require that
the QM department is responsible to review and approve production procedures, and any changes to them, most reports,
procedures, and controls, deemed necessary to assure the quality of the process and product.

Finally, the QM department must have adequate laboratory facilities and properly trained and experienced people to
effectively carry out their responsibilities.

F.   The Regulatory Authorities

Health authorities in every country regulate drug products.
In most countries, these regulations also include APIs. These
cGMP regulations require that a drug must meet all predefined quality specifications and be produced from a documented validated process. Further, if the drug, or API, is not
produced and controlled according to the established process,
then the drug is considered adulterated, and therefore not fit
for use or sale. The regulations address every aspect of drug
product manufacture, and essentially require that the producer has documented evidence of proof of control over any
aspect that might affect product quality.

The regulators were deliberate in their use of the word
"current" when the cGMPs were promulgated. This qualifier
enables the agencies to continuously require that manufacturers maintain their facilities and processes at the state of
the art, thereby always assuring the public that drug
products are as safe and effective as possible.

G.   The Regulations

The production of APIs is regulated in most countries. The
ICH-harmonized tripartite guideline Q7A entitled as *Good*

Manufacturing Practice Guide for APIs was recommended for adoption at Step 4 of the ICH process on the 10th of November 2000. This document was adopted by the following agencies denoting its widespread acceptance:

- European Union (EU) adopted by CPMP, November 2000, issued as CPMP/ICH/1935/00
- Japanese MHLW adopted November 2nd, 2001 MSB notification NO. 1200
- United States FDA published in the Federal Register, Vol. 66, No 186, September 25th, 2001, pages 49028–49029.

The production process and all tests and controls must be approved by the regulating government in which APIs will be used, and the facilities and systems in which they are produced must meet the manufacturing standards set down by the governing body. Thus, the quality of APIs is based on two components: meeting final quality specifications and being produced according to the regulated, approved process in a facility compliant with the appropriate manufacturing standards. It is important to note that both criteria must be met: final specifications and compliance to manufacturing standards. These two components will be dealt with separately in this chapter. It is also important to note that the approach toward quality described in this chapter should apply to any API regardless of the country in which it will be used or sold, or whether or not it will be a regulated item. The approach to quality, described in this chapter, is based on sound scientific principles, good QM principles, and applies to any API. In fact, these principles apply to the manufacture of any chemical that requires a high assurance of quality.

This chapter will deal with the chemical synthesis of APIs. However, all the principles and regulations also apply to other means of preparation, such as fermentation routes or extraction from natural sources.

Finally, since it is assumed throughout this chapter that the API will be subject to regulatory requirements, reference will be made to the regulations. If the reader is dealing with an unregulated item, such reference may be ignored, but the scientific principles on which the regulation is based should be seriously considered.

II. DEFINING AND ASSURING THE QUALITY OF THE ACTIVE PHARMACEUTICAL INGREDIENT

This section of the chapter addresses how to:

- define the necessary quality attributes
- test for them,
- design them into the process, and
- validate the process to assure consistent production.

As APIs are regulated articles, their quality is determined not only by satisfactory test results, but also the assurance that the process was conducted according to a validated process.

A. Defining the API Quality

The API must have its final chemical purity and impurity and its final physical attributes specified; some articles also require microbiological analyses to be determined, depending on the final dosage form and the manufacturing process involved. These attributes are established to assure an API will perform satisfactorily in the pharmaceutical manufacturing process and will result in a final dosage form; i.e., the drug product that will meet its initial release specifications and final stability requirements. The chemical purity minimum is usually set at 98% to assure proper dosing in the drug product and to assure a minimal amount of impurities. The physical parameters should be established with knowledge of the pharmaceutical process and the ultimate final dosage form. Other attributes usually include color of the solid form and/or a solution, melting point, specific rotation if optically active, crystal morphology, and so forth. A list of typical API specifications is provided in Appendix A along with the rationale for each one.

When setting API physical attribute specifications, the most important aspect to consider is its use in the pharmaceutical process; namely, whether it will be wetted for granulation, dissolved for solution, dry blended, and so on, and the type of drug product to be made: tablets, capsules, solutions, sterile or non sterile, or other. It is also important to know how the drug product will be used by the patient; for example, if it will be used as a powder blended with other excipients, careful consideration should be given to rate of dissolution and the eventual color of solution (for aesthetic reasons) when dissolved by the patient (or healthcare giver) prior to use. For this reason, final API specifications are always defined with the cooperation of the pharmaceutical development area. The quality assurance function approves final API quality standards, taking into consideration all requirements: process related, governmental, and customer.

B. Testing the API for Its Defined Attributes

Each quality attribute required of the API must have a sound and proven test procedure. In regulatory compliance terms, this means the test must be validated; that is, to have documented proof that it performs reliably, is indicative of the attribute under question, and is not biased by interfering components. There are eight specific components of a validated test, and for an excellent treatise on this, the reader is referred to the current USP or the ICH guidance on analytical test validation. Most regulatory authorities require a test for all significant API quality attributes on each lot produced. In nearly all cases, the pharmaceutical manufacturer requires a certificate of analysis (C of A(documenting the results obtained on each lot, as well as a statement from the quality office that the batch met its established quality criteria.

C. Designing Quality into the Process

As described above, the pharmaceutical manufacturing process and end use of the drug product dosage form are the basis for establishing the limits of chemical purity and physical attributes. Having predefined these attributes, the synthetic chemist and chemical engineer have the task of *designing quality into the process*; thereby assuring every lot will meet its criteria. This is perhaps the most significant aspect of chemical process validation and a cornerstone of most regulatory requirements for quality assurance. After the chemical process is developed, a technical document, which explains how and why certain reagents, steps, controls, etc. were chosen in order to build quality into the product, should be prepared. When the manufacturing team takes on the commercial implementation of the process, and goes through the formal manufacturing validation process, they should rely heavily on this technical document to prove the quality of the final API. As stated in the introduction, quality is designed into the process not for regulatory purposes, but because it makes good manufacturing and business sense to do so. Manufacturers want a process that safely and reliably delivers high yield and quality for economic and environmental reasons.

One should begin the approach to designing quality into the API, with the concept of designing a perfect system. Keep in mind that all the safety, environmental, and economic reasons for developing a perfect chemical synthesis are precisely consistent with the goal of designing quality into the process, and very well serve all regulatory process validation and control requirements. If one imagines a perfect process, there will be no toxic emissions about which to be concerned, no safety concerns or need for special safety controls, and the yield of each step will be 100% of the desired intermediate, stereo isomer, and end product. Such a process would be free of any impurities and would assay for 100% purity. The next challenge is to design the synthesis so that each step can be precisely controlled to always provide the same end result. The design work requires a complete understanding of the chemical reactions in the synthetic process under development. Then a clever design can be developed to eliminate any undesirable side reactions. In some instances, this can be achieved by sophisticated use of functional group protecting agents, and in other instances by changing the sequence of functional group introduction onto the end product building block and sometimes by simple careful control over reaction parameters. Once the process has been perfectly designed, developed, and controlled, the last concern is over the control of quality and reliability of the raw materials, proper functioning of equipment, and error-free operations by personnel. With the vision of a perfect system in mind, one can imagine how the API quality would be perfect and consistent.

D. Validation of the Process

This aspect of the regulations is perfectly aligned with business interests. The regulations require that a chemical manufacturing process be validated, which the author personally defines as *proof of knowledge of control*.

While the term "validation" has various definitions in several different regulations (cGMPs), all essentially mean or imply "proof of knowledge of control." In essence, the validation of the process is the description of the process after all development work is completed, with the elaboration of the proof of synthetic pathway, controls over process conditions, and finally, sound analytical proof of quality from samples obtained during actual manufacturing campaigns in the plant. Critical process parameters such as time, temperature, and mixing conditions should be defined, controlled, and monitored. The kinetics of the synthetic pathway is documented in a process manual. The establishment of a process manual for each API is the foundation of process validation. In this manual, one describes *proof* of the *knowledge* of the process and the *controls* necessary for consistent results. Hence, the scientific design process to build the perfect process requires full *knowledge* of the chemistry of the process. That knowledge is described in the chemical pathway from raw materials to the final API. The scientific evidence, such as intermediate structure elucidation, spectrographic analysis (IR, Near IR, mass spec, UV, NMR, C13NMR, etc.), and the proposed chemical mechanism for each transformation, serves as the *proof* of that knowledge. Finally, during the course of the process development, full knowledge is gained concerning those parameters and conditions that affect the kinetics, yield, and purity of each step. Experiments to optimize each step for purity and yield lead the process engineer to describe the necessary controls and conditions. These controls are described in a process manual and are used in the scale-up work and ultimate full-scale operation in the chemical plant.

E. Reality

We realize that the perfect synthetic process will, in all likelihood, be too elusive. Eventually, we must make the decision to focus our resources on the best process available after thorough development work yields a sound and reliable process. Each synthetic challenge represents reality of the business of API manufacturing, and so at some point, the feasibility of further studies vs. commercializing what has been achieved to date must be evaluated on a risk (loosing precious time in the market) to reward (achieving a superior process) basis. It is sufficient to say here that to ensure quality of the final API, the development of the process provides the necessary information to design in-process controls needed to monitor the progress of each step. These controls are the chemical and physical monitors that inform the operator that the synthesis is proceeding according to the original design. They are used also to inform the operator when the reaction is complete and when the next step may occur. In many cases, especially when the process is well defined and designed, including the quality of starting materials and reagents, a good control is simply the use of time, based on a knowledge of the kinetics of the reaction.

In-process controls should always be "in the process," that is, "on-line," and not requiring a sample to be withdrawn and sent to a laboratory for testing and evaluation. Under some conditions, it may be necessary to take samples, but this should be avoided whenever practical.

In-process controls are probes, or monitors, inserted into the reaction vessel, or the gauges that measure and record pressure and temperature of vapors above the reaction medium. The attributes that are measured include a wide variety relevant to the specific chemistry taking place. Properly established tests, for example, infrared or ultraviolet analysis, can predict the end point of a reaction by following the disappearance of a functional group on a reagent, or the formation of one on the molecule being produced. Monitoring the presence of any side products, such as water or gases, will signal the end point of the reaction when their theoretical yield is obtained.

III. THE REGULATIONS FOR QUALITY

A. Introduction: The Emergence of Specific Regulations for APIs

The active ingredient of a pharmaceutical product must meet two distinct sets of criteria before it can be used for producing a drug product suitable for sale in most countries around the world. One set of criteria is the product specifications, addressed in Part II. The other set is the assurance that the product is produced according to cGMP, that is, the cGMPs prevalent in the regulated market in which the drug product will be sold. Most countries have approved and enforced regulations for drug products; there were few with specific regulations for the APIs used therein. With the development and subsequent adoption of the ICH Q7A Guide, by the EU, MHLW and US FDA, a consistent approach to cGMPs for the manufacture of APIs is now achievable.

The section is written as if the regulations are in force throughout the world. This position is valid given the adoption of the ICH Q7A Guide.

One final introductory comment before beginning a review of the ICH guidance: when describing the "cGMPs," they are always prefaced by the adjective "current". Q7A acknowledges the equivalence of the terms "cGMPs" and "good manufacturing practices". The equivalence of the terms is deliberate. It requires that manufacturers continuously apply the current state of technology and practices when developing new drugs. In certain special cases, manufacturers will also be compelled to apply the new technology to older APIs and the processes, facilities, etc., whenever such application will play a significant role in assuring, or advancing, the end product quality. Furthermore, since the "current" is part of the guidance, manufacturers need to be aware of such advances and make the necessary changes to their systems and facilities to remain compliant. Hence, the guidance can be thought of as always being updated without return to the regulating bodies for approval.

This part of the chapter follows the format of the ICH Guide as finalized by ICH in November 2000 (available from www.ich.org—see "Quality," then "Q7A"; or www.fda.gov). However, it is not intended that this will represent a summary of the guidance. Instead, this text offers practical insight into the reasons and meanings of certain aspects of the requirements. Review and reference should that be necessary. The reason for selecting the ICH guide is due to its widespread adoption, its comprehensive approach and its high quality as a reference document.

The ICH guidance is laid out in the following format to demonstrate the scope and extent of their influence on the entire manufacturing process (note we are following the ICH numbering format), each of these ICH sections are discussed below:

1. Introduction
2. QM
3. Personnel
4. Buildings and facilities
5. Process equipment
6. Documents and records
7. Materials management
8. Production and in-process controls
9. Packaging & identification labeling of APIs and intermediates
10. Storage & distribution
11. Laboratory controls
12. Validation
13. Change control
14. Rejection and reuse of materials
15. Complaints and recalls
16. Contract manufacturers (including laboratories)

17. Agents, brokers, traders, distributors, repackagers and relabelers
18. Specific guidance for APIs manufactured by cell culture/fermentation
19. APIs for use in clinical trials
20. Glossary

(Note: Sections 17–20 are not discussed within this chapter).

To assure the manufacture of an API meets the requirements described in the ICH guide is, indeed, a significant task. It requires that all the requirements are understood by all appropriate people dealing with the manufacturing process, that their understanding is proven through documented training records, and there are effective systems and procedures in place to assure that all appropriate steps, controls, tests, etc. are conducted, as described, in the product's New Drug Application, in the firm's Drug Master File, and in the documented standard operating procedures of a manufacturing facility.

Let us begin with an analysis of the ICH Q7A Guidelines, section by section.

1. ICH Q7A Section I: "Introduction"

This section describes the scope and application of the guidelines. It provides guidance as to when cGMPs should be applied to the manufacturing process. The grid from the Q7A Introduction (Fig. 1) demonstrations GMP applications. It shows the various types of API manufacturing technologies, for example, from "chemical manufacturing" on the top left of the grid through to "classical fermentation" on the bottom left. For each technology, moving left to right, the likely processing steps that might be used are mentioned. As the process moves closer to the final steps, the degree of cGMP requirements increases. At some logical point, the heads of manufacturing and quality decide on the "starting materials" that will reliably produce the API. These starting materials must be very well characterized, and always be tested for conformance to predefined attributes before use in the API process. Admittedly, which chemicals are defined as "starting

Type of Manufacturing	Application of this Guide to steps (shown in gray) used in this type of manufacturing				
Chemical Manufacturing	Production of the API Starting Material	Introduction of the API Starting Material into process	Production of Intermediate(s)	Isolation and purification	Physical processing, and packaging
API derived from animal sources	Collection of organ, fluid, or tissue	Cutting, mixing, and/or initial processing	Introduction of the API Starting Material into process	Isolation and purification	Physical processing, and packaging
API extracted from plant sources	Collection of plants	Cutting and initial extraction(s)	Introduction of the API Starting Material into process	Isolation and purification	Physical processing, and packaging
Herbal extracts used as API	Collection of plants	Cutting and initial extraction		Further extraction	Physical processing, and packaging
API consisting of comminuted or powdered herbs	Collection of plants and/or cultivation and harvesting	Cutting/ comminuting			Physical processing, and packaging
Biotechnology: fermentation/ cell culture	Establishment of master cell bank and working cell bank	Maintenance of working cell bank	Cell culture and/or fermentation	Isolation and purification	Physical processing, and packaging
"Classical" Fermentation to produce an API	Establishment of cell bank	Maintenance of the cell bank	Introduction of the cells into fermentation	Isolation and purification	Physical processing, and packaging

Increasing GMP requirements

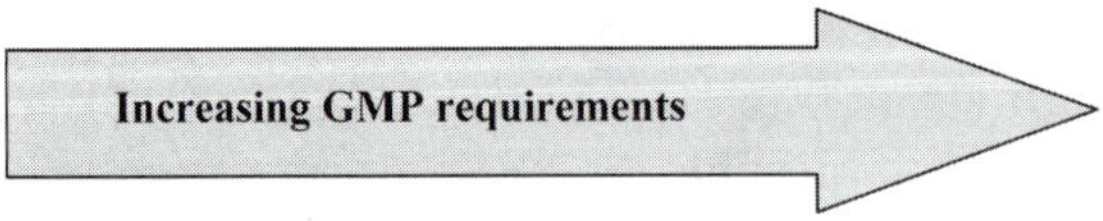

Figure 1

materials" can be debated. Logically, if a reagent will become part of the final molecule, it is a very good candidate to be chosen as one of the "starting materials." From this point forward, the cGMPs applied to the process increase in their stringency. Final packaging of the APIs will be very well prescribed in both the environmental requirements and in the labeling controls applied.

a. Compliance Requirements

Defining the point at which cGMPs first come into effect for the API (or intermediate) being produced is the essence of this section. From that point forward, the strategy and the requirements for consistently producing APIs in conformance with cGMPs are developed.

2. ICH Q7A Section 2: "Quality Management"

Beyond the principle that quality is the responsibility of all personnel, there are specific responsibilities, which must be carried out by the quality unit. The challenge is to achieve a balance between executing the necessary activities and not allowing the abdication of other departments toward quality. The quality unit most often becomes a focal point for all quality related matters, functioning as a technical consultant in quality and compliance. It is imperative that the quality unit be independent of the manufacturing operations in order to achieve an objective perspective. Some of the basic quality responsibilities include review and approval of documents (specifications, test methods), written SOPs for all departments, records (batch records and log books), deviations and their investigations/resolutions, and finally the release of the product to market.

Internal auditing (also termed self-assessment and self-inspection) is also a cornerstone of QM. Knowledge of the plant and its systems should be used to determine the annual schedule for conducting audits. Once executed, the information contained in an internal audit is useful not only for the department being audited to improve their operations, but also for the plant on a larger scale to know about potential quality issues in advance. Senior management must be made aware of the issues found in internal audits since they are responsible for setting the strategy for the plant and can allocate resources to correct any deficiencies.

In addition to internal audits, which are carried out on manufacturing processes, a review of the product as it is manufactured, and its ability to meet specifications both initially and over time, can yield information for improving processes and the product itself. These reviews are typically captured in an annual report known as the product quality review, annual product report or product quality history. Each plant defines the scope of the report, however but minimally the results of all tests (at the time of release and when tested subsequently), the review of deviations and difficulties in its manufacture, complaints received from customers, and especially looking for trends that may not be apparent on a day-to-day basis are the foundation of any product review.

a. Compliance Requirements

A well-defined internal auditing program, comprehensive in scope with a good communication plan, can help identify and prevent systemic problems in the operation. The internal auditors must not only have a thorough knowledge of the operations and the regulations, but also the best means of achieving conformance without creating unnecessary bureaucracy. Combined with annual product quality reviews, opportunities for quality improvement should be identified both for the products and the processes.

3. ICH Q7A Section 3: "Personnel"

No matter what API is made, there are always people involved in the process. It has been said by numerous CEOs, "people are our greatest resource." Each manufacturing operation needs the right level of personnel both in terms of number and qualification for each job in the operation. This presupposes that each job has a well-defined job description complete with training requirements and the demonstration of the proficiency of the necessary activities. While this has been common in the laboratories, companies are now expanding the concept of training and qualification to the informal programs common in our industry. The use of "mentors" exists in most manufacturing operations but few ask, "how are these mentors chosen, what are the skills that make one person an excellent mentor and the other a poor one, what training is needed to make the mentoring effective and efficient?" The recognition of mentoring as a training activity will increase the knowledge base resident in your work force.

Since people are involved in the manufacture of APIs, their hygiene becomes important. The firm must provide adequate toilet, cafeteria, and changing/locker facilities. This achieves not only protection of the personnel, but also of the product.

It is common practice to employ consultants or contractors for limited times on short-term specific projects. Consultants or contractors must follow the same requirements for personnel hygiene and must have the training necessary to perform their duties.

a. Compliance Requirements

All personnel involved in the manufacture of APIs and intermediates must have the necessary training, education, and skills to perform their activities in a consistent manner. This must be documented not only for regular employees but also for contractors/consultants as well. Proper hygiene and sanitation protecting the product and the personnel must be part of normal operating conditions at the site.

4. ICH Q7A Section 4: "Buildings and Facilities"

In planning a facility, there are always drawings of the facility; the equipment, the utilities, the material flow, and personnel flow. This is necessary to assure that in the design of a facility, adequate space is provided for the material and personnel to flow smoothly, and for the prevention of mix-ups and contamination.

Utilities can have direct impact on the quality of the API. The utility can either be required to assure consistent environmental parameters, as evidenced by heating, ventilation, and air conditioning (HVAC) or by providing materials that come in direct contact with the product (compressed gases). Inconsistency in either of these functions can result in inconsistent quality of the product and/or process. As such, all utilities having a direct impact must be qualified through rigorous testing to show it will consistently perform as expected. In addition, these utilities have to be monitored to assure they are continuing to function as necessary.

Water can be used for a variety of purposes within the manufacturing of APIs. Because of the many differing uses, such as a cleaning material or a raw material in the manufacturing process, how the water is used will determine the specifications to be developed. Beyond how the water is used, the product it is used in will also determine the specifications. At a minimum, process water must meet World Health Organization's (WHO) guideline for drinking (potable) water quality. If this water is then treated to yield a water of a predefined quality, that process must be validated and the water produced monitored. This can include specifications for the microbial content or endotoxins in the water for an API to be used in a sterile product.

Containment considerations must also be built into the design of the facility and their utilities in order to protect the workers from adverse exposure to the API, or its intermediates. Certain APIs are highly toxic or potentially deleterious materials, such as penicillin and cephalosporins. The side effects from these drugs can be life threatening. It may be necessary to develop separate and dedicated facilities for the manufacture of these types of compounds.

Lighting should be sufficient for all personnel to perform their activities without eyestrain. Certain APIs may need to be protected from light. In those cases, the lighting may have to be designed taking into account the product's requirements and the personnel's needs.

Every operation will generate waste. In the case of the manufacturing of APIs, this waste can cause contamination throughout the manufacturing facility if not removed in a sanitary and safe manner.

How a facility is maintained is one of the first indications of the quality culture at the site. A high-quality API requires a facility that is clean, free of pests and has utilities that function reliably. The cleaning agents or other materials used to maintain the facility of equipment must be known not to contaminate the product in its usage. Reliability of the utilities can be assured by a well-defined and executed preventive maintenance program.

a. Compliance Requirements

Engineering drawings detailing the facilities layout, equipment location, room usage, material flow, personnel flow, and utilities must be maintained to reflect accurately the building and facility. Procedures for the maintenance and monitoring of all the utilities that can impact product quality must be written and approved. Utilities producing materials that are used in the actual manufacturing operations must have specifications for those materials. Safety of the personnel and the prevention of the product contamination must be thought of in advance and captured in standard operating procedures specifying sanitation, containment, and refuse disposal practices. The maintenance of the facility, the building, and the utilities must be defined in operating procedures using materials that will not adversely affect the product.

5. ICH Q7A Section 5: "Process Equipment"

For each manufacturing operation, there will be equipment either fixed or mobile to be used in the process. Sometimes the equipment may be placed outside of the building itself. Equipment can be either closed (preferable) or open. If the equipment is not closed, there must be special attention to prevent contamination of the product. In either case, the equipment must be constructed of the right materials to assure that it can be easily cleaned and maintained. It must protect the product it is manufacturing and not affect the quality of the product. Identification of the equipment (and processing lines) is necessary to assure traceability of the product to the equipment used. Not all equipment will need to be identified; each facility must define which pieces of equipment are considered major in their own manufacturing processes. The use of lubricants or other manufacturing aides is often necessary, however, when used, these materials must not contact the API or alter the quality of the API.

A clean, well-maintained building is only half of the picture. A clean facility with broken or rusty equipment is just as indicative of a poor quality culture as is the reverse. The equipment in the building must be maintained in a good state of repair and cleanliness. The expectation must be that once processing starts, all the necessary equipment will function properly.

Cleaning materials must be chosen not only to clean the equipment, but also to leave no residues, which may affect the quality of subsequent batches. How long a piece of equipment remains "clean" after cleaning is completed is an important factor to be considered in designing the cleaning program. Also part of a comprehensive cleaning program is whether reduced cleaning can be used for campaign manufacturing (subsequent batches are produced of the identical API).

Manufacturing equipment often provides data, information and output, which can be used to determine the acceptability of the processes and/or the product itself. We must be able to rely on this data as correct and accurate. Therefore, the analytical components of this equipment must be calibrated using standards whose authenticity is assured. The time in between calibrations will be based on the reliability of the equipment and the criticality of the data. When calibration requirements are not met, decisions based on the data may be erroneous and must be re-evaluated. Instruments, which do not meet calibration requirements, should not be used and there must be an investigation to assess the impact on batches produced in that equipment.

Computerized systems are specialized pieces of equipment which must meet all the requirements as other pieces of equipment. A logical application of the cGMPs should be applied in conjunction with special requirements such as 211 CFR Part 11. Validation is required, however, the depth and scope of the validation is dependent of the computerized application. The degree of validation may be dependent on the source of the computer system: commercially available software requires little validation, while software developed for a specific manufacturing step will require extensive validation. Once validated, the computer system must be maintained in a state of control. After all, it is still a piece of equipment, albeit highly specialized.

a. Compliance Requirements

Procedures for the cleaning, maintenance, and operation of each piece of major equipment must be developed and approved by the quality organization. There also needs to be engineering drawings of the equipment along with the maintenance of a revision history of these drawings. Many firms see engineering drawings analogous to documents and require quality approval of changes.

6. ICH Q7A Section 6: "Documents and Records"

In a sense, our industry prepares two products: the API that goes out the door and the paperwork which records how this batch was made. The paperwork consists of documents such as procedures, specifications, methods, and manufacturing instructions. These documents are prepared and revised by a formal change control process; the quality unit must review and approve them. The issuance of these documents to the operators, technicians, and other persons using these documents must be done in a controlled manner, assuring they have the most recent version of the approved document. Typically, the quality unit has a documentation center which will store the superseded documents and the approved master copies. The storage of documents must be defined in a records retention policy consistent with cGMP requirements (at least 1 year after the expiry of the batch or for APIs with retest dates, 3 years after the complete distribution of the batch) and any legal requirements of the firm.

The data detailing the actual manufacturing conditions (temperatures, lot numbers of starting materials, etc.) is recorded on a batch record. Data detailing equipment usage and conditions are similarly considered to be a cGMP record. Changes to cGMP records must show the person making the change and the date the change was made. Many firms also annotate the correction with the reason for the correction.

Only accepted raw materials may be used in the production of an API. This sets into motion a requirement for proper documentation surrounding the receipt, testing, acceptance, and release to manufacturing. Imagine that a raw material is found to be defective after it has been used in a product.

The firm would have to trace (using the lot number of the supplier) to determine which batches this defective raw material has been used in. The records then must include the supplier's lot number, a unique lot number assigned to the materials, date of receipt of the material, and records of the testing and release of the material.

Manufacturing Instructions are documents detailing the process and the controls necessary to produce uniform quality batches, time and again. In a sense, the manufacturing instructions are the recipes for making the API.

Specific requirements for these records include:

- Name of the API and any unique identifying code.
- List of raw materials and the quantity or ratio of materials to be used. Specific calculations must be included as part of the specific batch record.
- List of equipment to be used.
- Specific production instructions in the correct sequence with all necessary control parameters, time limits, yield ranges and in-process control testing and specifications.
- Instructions for storage.

There should be a master batch record document from which each individual batch record is generated. This insures that each lot made is manufactured following the same recipe. When the lot specific batch record is issued, there must be a check to assure that the issued batch record is specific for a single unique batch number and the instructions are identical to the master batch record document. The lot specific record will have the signature of those individuals who performed critical processing steps in the manufacturing process and those who performed any in-process testing.

Eventually, irregularities will occur. Chemical manufacturing processes are a complex interaction of reagents, solvents, machines, and people. Machines can break down and people can make mistakes. In-process materials, or products do not always meet specifications. Each firm must have a system in place to identify and assess the impact of these deviations on product quality and process robustness. When deviations occur, they must be addressed as part of the batch record. Prior to the release of the batch, the quality unit must approve the deviations and how they have been resolved.

As the batch record details the executed processing steps, there are records for the laboratory testing for each batch. These requirements are completely consistent with the cGMPs for laboratory records for finished pharmaceuticals. Samples are taken to determine the acceptability of a material; the traceability of a sample to the portion of the lot where it came from must be known and recorded. We must know definitely what tests were performed on the sample (test method including revision date), what equipment was used to generate the data, who performed the test, who reviewed the data (a second person must review the data) in addition to the acceptability of the test results themselves.

When the manufacturing and the testing activities are completed, a review of the critical process steps and the noncritical process steps must be performed. The quality unit must review the critical process steps; the noncritical process steps can be reviewed by a different unit following procedures that have been approved by the quality unit.

a. Compliance Requirements

Any aspect of producing an API, from the receipt of all incoming materials from outside vendors right through to the last distribution of released material, must be appropriately recorded. Systems must be in place so that each specific lot of intermediate or API is reviewed and approved by QC before it is released. Effective systems need to be in place to both detect an unexpected result, and then to investigate it. Batch production records, or a sound sampling thereof, as well as all other quality related records such as stability data complaints and so forth should be reviewed at least annually to ensure that in-process controls, procedures, and final product specifications are adequate and tightened where appropriate.

7. ICH Q7A Section 7: "Materials Management"

There must be a comprehensive system with procedures defining the receipt, identification, quarantine, storage, handling, sampling, testing, and disposition of materials. There must be predetermined specifications for materials purchased from a supplier, which has been approved by the quality unit. This presupposes that there is a system for approving suppliers and the materials they supply. Critical raw materials must be identified and changes to any specifications of critical raw materials must be handled under change control.

A manufacturing plant will receive several different types of incoming materials, some require the strictest degree of controls (raw materials and intermediates, for example) and others can be handled with little more than good accounting practices (office supplies, etc.). For each material to be used in the manufacturing process, there must be a unique lot number assigned to the goods to assure complete traceability to the supplier and the shipment. Shipping conditions can affect materials even during a short time; shipments of the same supplier lot number may require a different receiving lot number. Upon receipt, the containers should be checked for conformance to the labeling and be free from tampering or damage, which may cause contamination of the API. Before any material is used, the quality unit must formally release the material.

The use of nondedicated tankers requires an additional level of assurance that any potential contamination is prevented. This can be assured by audits of the suppliers and/ or a certificate of cleaning supplied by the supplier and/or testing for impurities by the receiving firm.

Many firms rely on a C of A supplied by the supplier in lieu of actually performing required testing. This becomes a more proactive means of assuring the quality of the material. Acceptance of a C of A is possible after a partnership is established with the supplier through a formal qualification and evaluation of the supplier's capabilities and reliability. Typically the process involves an initial questionnaire, followed by an audit by trained auditors and purchasing representatives. This helps determine if the supplier is qualified to produce the material consistently and in accordance with the firm's expectations. The material then needs to be approved for use in the manufacturing process; this is typically done with three distinct batches produced by the supplier. The goal is to determine if the supplier's material performs reliably in the firm's manufacturing process. The entire process of qualifying the vendor and the material they produce must be repeated on a periodic basis. For hazardous or highly toxic raw materials, full acceptance of the C of A may be warranted pending a documented rationale.

It would be easy if all materials were received as a single lot, in a single container; however, it is often the case that the vendor will use multiple containers—and sometime different lots, typically to facilitate his own handling and shipping efforts. Each container must be inspected at the time of receipt; however, the contents of each container need not be sampled and tested to determine the material's acceptability. The use of statistical sampling plans applied to each lot separately can help reduce the burden of sampling and testing while still yielding a result representative of the batch as a whole. These sampling plans must take into account the number of containers received and the criticality of the material.

While these materials are held, either prior to release in inventory or in the manufacturing process itself, they must be held and handled so as not to contaminate the material itself, or other materials stored in the area. For example, heat sensitive materials may need to be stored in a cool, controlled (i.e., data-recorded) location. It is quite common in API manufacturing to store materials outdoors. This can be accomplished for specific materials as long as the requirements stated above are met. When brought into the manufacturing environment, the containers may require an additional cleaning.

If a material is determined to be unfit for use and rejected, special storage and handling requirements must be met. It must be stored in such a manner that it cannot be used inadvertently in the manufacturing process. Many firms have designated locked cages to store rejected materials.

a. Compliance Requirements

The receipt, storage, and handling of materials must be performed in such a manner that there is complete traceability of the material to the supplier. Materials may be received against a C of A supplied by a qualified supplier in lieu of full testing. For this to happen there must be a formal supplier qualification process. For materials that are sampled and tested for release, there must be predetermined statistically valid sampling plans and acceptance criteria. Materials must be segregated or otherwise stored to prevent their use, until the formal release by the quality unit is granted. The materials must be stored in such a manner as to not compromise their quality attributes. Any rejected materials must not be used and must be stored in a separate area to prevent such an error. Rejected material should be returned to the vendor, or otherwise properly destroyed.

8. ICH Q7A Section 8: "Production and In-Process Controls"

Earlier, the need for a master batch record document was discussed. This document describes the manufacturing instructions necessary to consistently produce batches of APIs that meet predetermined specifications. There will also be ancillary procedures, which will define all the conditions and their control parameters necessary to assure consistency from batch to batch. Isolated materials should be labeled at each step in the process. This is true not only for the raw materials but also as in-process materials are generated and isolated, the material's name, lot number, and its status should be clearly labeled. As stated earlier, major pieces of equipment should be clearly labeled with a unique identification number and its status (cleaned and ready to be used, to be cleaned, or in use). If a processing step is determined to be a critical processing step, it may require witnessing of its completion by a second person, with the witnessing documented on the batch record.

Processing steps should have time limits, either a step must be completed within a certain time (mix for 2 hr) or an in-process material may be held for a specified amount of time. Deviations from these time limits must be addressed with a formal investigation as to the affect they may have on the quality of the product. The results and conclusions of the investigation must be documented.

The manufacturing process should be reviewed for those points where in-process testing can minimize variability in the process thus achieving greater consistency in the yield and quality of the product. This review should ideally take place during development runs but information from validation batches and annual product quality reviews can also be used. They can be valuable in determining which areas of the process or product should be more carefully monitored or if process specifications should be changed. Types of inprocess control tests include temperature of the process, pressure of equipment, the color of solutions, pH, loss on drying, etc. In-process control testing and specifications should be defined in documents approved by the quality unit.

Each intermediate must meet its quality requirements before further processing. Even if two or more batches of intermediates will be blended prior to the next step, they must each meet their respective quality requirements.

Blending is a process somewhat unique to the manufacture of APIs. It is an accepted practice to blend batches of the final API as long as each of the individual batches meets the predetermined specifications prior to the blending. After the blending is complete, there must be a final test to assure that the final blended lot is acceptable as well.

A sample of the final lot, blended or otherwise, should be taken and stored for future stability testing if necessary.

Campaign manufacturing is common in pharmaceutical operations. If successive batches are being made of the same API, it is acceptable for residual materials to be carried from one batch into another, as long as there is adequate control. Adequate control will need to be determined for each API produced but minimally there must be assurance that degradants, microbial contaminants, or other sources of contamination are not carried from one batch to another.

All operations must be conducted in such a manner that contamination is minimized. This is especially true after the purification steps in the manufacturing process.

In addition to preventing contamination of the API, the safety of the operators must be addressed as well. If the materials may be injurious to the operators and/or the environment, how best to assure there is no harm to either must be addressed as part of the development of the manufacturing process. It may be necessary, in some cases, for the process to be carried out using dedicated equipment or in dedicated and controlled environments.

a. Compliance Requirements

Clearly, this section requires a large volume of documentation, with the focus being proof that each batch has been produced according to the original design. Each batch is individually monitored, and its manufacturing history is recorded in a production batch record. In-process controls should be evaluated after adequate experience is gained in full-scale production. Test limits must be changed, i.e., tightened, if justified by historical results. The QC unit must review and approve all changes to production records, control procedures and test procedures and/or limits. Any deviation from established procedures, whether planned or not, must be investigated for cause, and documented for corrective action.

9. ICH Q7A Section 9: "Packaging and Identification Labeling of APIs and Intermediates"

This section has four subsections, which state quite specific yet reasonable requirements for packaging materials, labeling issuance and control, and packaging and labeling operations of APIs. They are remarkable in their similarity to the requirements for drug products. They require tight control over the receipt, testing, release, storage and use of containers, and labels. Particular care must be taken to avoid mixups of labels, and separate storage areas should be provided for all different labels. Further, access to the area should be restricted to only certain authorized personnel. The labeling operations require the same assurance that the labeling facilities are separate from other activity and that they are adequately cleaned prior to use. All preparatory work must be documented in written procedures. Packaging and labeling operations should be recorded in the manufacturing batch record for each lot of API.

The label must bear the usual descriptive information about the product and include its distinct lot number. For intermediate or APIs with an expiry date, the date must be indicated on the label and on the C of A. For intermediates or APIs with a retest date, the retest date should be indicated on the label and/or C of A. Naturally, an effective system must be in place to assure no materials beyond their expiry period are used, and all those requiring retesting are completed before use beyond the controlled time period.

As in other areas of API manufacture, contact surfaces must not be reactive, absorptive, and so forth with the API so as to alter its quality.

a. Compliance Requirements

Well-documented systems must be in place to handle the receipt, testing, release, and use of containers and labels, similar to those control procedures used for raw materials. Packaging and labeling operations must be conducted in separate areas to avoid contamination and mix-ups with other ongoing activity. Inventory management of labels must be practiced, with accountability of all used and remaining labels kept up to date for each lot of labels. Expiry dates or retest dates must be based on analytical evidence obtained under the intended storage conditions, and each specific date for each lot must appear of all labels used to package each lot.

10. ICH Q7A Section 10: "Storage and Distribution"

This brief section contains only two parts. It directs that the warehousing procedures and distribution procedures must be written and, of course, be consistent with the intended storage conditions for which stability data exist for the material. Materials should be held in a quarantine condition until released by the quality unit. This status control may be a physical separation with appropriate labels or, ideally, through the use of electronic control systems. The distribution history of each lot must be maintained for traceability in the event a recall is necessary.

a. Compliance Requirements

Have written procedures to describe the handling of the materials. Ensure that a tight system is used to prevent use of materials before release by QC. Material should be distributed on first-in-first-out (FIFO) basis.

11. ICH Q7A Section 11: "Laboratory Controls"

This section is described in greater detail in Part III of this chapter.

12. ICH Q7A Section 12: "Validation"

This is a most important section of the cGMPs. As such, it warrants its own chapter in this book. Here are described the essential compliance aspects of validation.

Within the ICH Guide, validation is further divided into the areas of its disciplines:

- Validation policy
- Validation documentation
- Qualification (Equipment IQ/OQ)
- Process validation
- Periodic review of validated systems (Revalidation)
- Cleaning validation
- Validation of analytical methods

The validation policy is a high level document stating the approach a firm will use toward validation. The validation approach requires the development scientists and plant management to identify the critical parameters/attributes during the development stage and use that knowledge in the validation of the process.

For each validation activity, there is a validation protocol (a study design) written and approved in advance of the execution of validation work. The quality unit must review and approve the protocol, as other affected departments. The protocol lists the tests to be conducted along with the acceptance criteria; the tests are chosen to demonstrate the process is in a state of control.

Once the validation protocol is executed, the results of the tests are written into a formal report. Any deviations from the acceptance criteria must be addressed in the report, along with a conclusion about the impact on the consistency and reliability of the process.

Validation is a life cycle process, which has its roots in the development area.

- Design qualification: verification that the proposed design is suitable for intended use.
- Installation qualification: verification that insta-l lation complies with the approved design, manufacturer's recommendations, and/or user requirements.
- Operational qualification: verification that the equipment performs as intended throughout anticipated operating ranges.
- Performance qualification: verification that the equipment and/or process can perform according to preapproved specification consistently.

Just as there are phases in the validation lifecycle, there are three distinct approaches a firm can take toward the validation process.

1. Prospective validation is the preferred approach and is the most common. If other approaches are used, the firm should have documented rationale as to why they did not use a prospective validation approach. A prospective validation is a formal study that serves to prove the process will reliably yield an API to meet its predetermined quality attributes and all steps along the way are reliable in terms of quality and yield. Validation can best be defined as *proof of knowledge of control*. For new products, or changes to processes requiring a process validation, the number of runs must be commensurate with the complexity of the process, or the nature of the change under review. Three consecutive successful production batches are typically required; exceptions should be documented. Process validation must confirm the impurity profile of the API.

2. Concurrent validation can be used where a small number of API batches are made on an annual basis. Typically, there are three validation batches manufactured in a study and all three are held pending the results. In the case where the time period between the production of the first batch and third batch is extremely long, concurrent validation can be used. In this case, the first batch is released on its own merit, however, the process is not considered validated. Only when the full protocol requirements are met, both in terms of acceptance criteria and in the number of batches are the process considered validated. The use of a concurrent validation is very rare and only suitable under special circumstances.

3. Retrospective validation is very rare and must be used judiciously. This approach involves reviewing a large number of batches already produced at the plant to affirm the robustness and repeatability of a process. There are very specific assumptions that must be met before retrospective validation can even be considered. There must not have been any changes made to the process during the review time period. The process must be a well-understood and -characterized process with defined in-process tests and controls. There must not be any significant process failures or deviations during the time period. The impurity profile for the product must be well established. Even when all these conditions are met, the decision to use retrospective validation must be a last resort and the justification well documented with approval from the quality unit.

Whatever approach is used for the validation, the goal is to gain the *proof of knowledge of control* of a process. A study plan, called a protocol, is prepared describing the important parameters that need to be controlled in order to assure the API will meet its quality parameters and expected yields. To determine those important parameters, data, results, and reports from the research department are used. During the initial development of the process, the controlling parameters should have been discovered, including effective working ranges and targets for charge of components, raw materials, and operating conditions of time, temperature, pressure, mixing rate, and so on. Analytical methods used to evaluate each chemical and physical attribute are themselves first validated. In this way, the data generated in the validation are known to be true and accurate.

The protocol defines how the study will be done (the process, equipment, critical steps, and parameters), and who is responsible for its design, execution, analysis, and approval. The sampling activity is also well defined, describing the locations to be sampled, the sampling devices to be used, the quantities required and time point during processing when they should be taken. The protocol defines all the important process parameters to be studied and analyzed in order to demonstrate each significant step performs reliably in terms of quality and yield. The final approval is reserved for the quality function.

A successful validation study will demonstrate a reliable and robust process. To be able to reach a strong conclusion, there must be an adequate number of batches and tests to statistically demonstrate reliability and robustness.

As part of the lifecycle approach to validation, periodic evaluations of the processes and products will determine the periodicity of revalidation. Certain processes will require an annual revalidation. Revalidation also occurs when there are significant changes to a process or piece of equipment, which would "void" the original validation.

a. Cleaning Validation

After manufacturing is completed, the equipment should be cleaned and made ready for the next process. There needs to be a formal study, executed against a protocol, which demonstrates that the cleaning process used is effective to clean the equipment to a predetermined level of cleanliness. Cleaning validation is part of assuring that contamination and cross contamination are prevented. The protocol must include a description of the equipment to be cleaned, the materials for cleaning to be used, and the cleaning process. The sampling equipment, locations, and procedures must be defined. In addition to visual cleanliness, where analytical methods are used, these methods must be validated to appropriate levels of detection. The limits of detection must be

based on sound scientific reasoning. Cleaning is not only performed to remove chemical contaminants but microbiological
contaminants as well. The removal of microbes and endotoxins must be addressed in the protocol where appropriate.

The cleaning process must also be included when there is
an evaluation as to the necessity of revalidations.

Analytical methods must also be validated. The approach
used for method validation is consistent with the validation of
analytical methods for drug products. See USP monograph on
analytical validation procedures.

b. Compliance Requirements

Assure the critical steps and intermediate quality attributes are defined and based on scientific rationale, usually
from original research information. The person making those
decisions must be identified in the protocol. Once the protocol
is approved, it cannot be changed during the course of the
study. While the documentation of validation studies is a regulatory requirement, it serves the business aspects perfectly
because it captures the intellectual property of the firm. The
protocol should include ranges for operating parameters.
These should come from research information. They need
not be tested or challenged during the validation study in
full-scale equipment.

13.   ICH Q7A Section 13: "Change Control"

Having established a validated process, efforts must be implemented to assure that it stays in the validated state. Systems
need to be implemented to evaluate both planned and
unplanned changes to the process. This refers to any change
in materials, conditions, equipment used, and site of manufacture, scale, and so forth. All planned changes must be
described and evaluated before implemented. All concerned
departments are involved in this analysis; the final review
and approval is required of the quality unit.

Factors to be considered in evaluating the change should
include any reasonable aspect of the API or its intermediates
that may be affected. This must include attributes that are

not routinely tested, such as polymorphism, the emergence of new impurities, and the need for additional stability studies, for example. For this reason, chemical experts familiar with the science must be consulted.

Finally, an analysis of the impact of the change on any filed regulatory documents is necessary, as well as informing the pharmaceutical users of the API. Pharmaceutical manufacturers might have additional quality and performance criteria, which are unknown to the API manufacturer, and these criteria need to be assessed relative to the process change under review.

The completely analyzed and studied process change is then evaluated by the group of experts who designed the change. The final review and approval are again required of the quality unit.

a. Compliance Requirements

An effective communication system needs to be in place to ensure that the quality unit is informed and involved in planned changes. Unplanned changes are to be discovered through the periodic review of production records (see earlier section). Changes may be classified to their expected degree of impact, and studies can be modified accordingly. Scientific judgment must always be used in evaluating the changes. Systems must be in place to ensure that material under change review is not used for further processing until approved by the quality unit. The decisions about what to evaluate must be documented, as well as why no additional studies are deemed necessary (for example, why polymorphism will not be affected by the change).

14. ICH Q7A Section 14: "Rejection and Re-Use of Materials"

This section describes the requirements for rejection of materials, reprocessing and reworking, recovery and recycling of solvents in the process, and customer returns of materials.

If specifications are not met and/or if the material is not manufactured in accordance with cGMPs, these materials must be set aside, and quarantined until disposition. The disposition can be rejection, reworking, or reprocessing the material. Material is used here to denote incoming materials, intermediates and/or finished product (APIs). If reworking or reprocessing is determined to be in accordance with regulatory controls, the actual reworking or reprocessing must be conducted and recorded in a manner identical to that of the original manufacturing steps.

The manufacturing of APIs can be distinguished from finished pharmaceuticals in that reprocessing is a far more accepted practice for APIs. "Reprocessing" is defined as the return of an intermediate or an API back into the process and repeating a part of the manufacturing step. Types of reprocessing include either physical reprocessing, for example the repetition of a drying step, or extending a chemical step. However, if reprocessing is used routinely for any given step, at some point it becomes the normal process. At that point the reprocessing step(s) should be incorporated into the manufacturing process and batch documentation, not as a reprocessing step, but rather as a routine part of the process.

Reprocessing by chemical means involves the repeating of a chemical reaction. This is rarely appropriate since in repeating a chemical step, new impurities could be produced. A batch requiring chemical reprocessing should first be considered for destruction before salvaging through reprocessing occurs.

Reprocessing by physical means involves the repetition of a step such as a recrystallization or remilling already routinely performed in the validated process. If the routine process does not include such a step, then it is not reprocessing but rather reworking. The actions to be taken in a reprocessing must be documented with a documented rationale for the reprocessing. Reprocessed materials must be evaluated to determine if additional testing is warranted.

"Reworking" is distinguished from reprocessing in that reworking is the use of a new step, or steps, not part of the routine process. While reprocessing does not require a new and separate process validation, reworking must have its own validation. Reworking requires the approval of the quality assurance unit. Reworking often requires a notification to the governing regulatory authorities, and typically requires their approval of the change before putting it into use.

Recovery and recycling of reactants, intermediates, or API's, in order to be used again, are considered acceptable. However, the use of these materials must be done using validated and documented procedures. If a recovered solvent is to be used in a different process (e.g., to produce a different API), there must be adequate validation and documentation to assure that it can be used without concern of cross contamination. Recovered solvents must meet predetermined specifications. Where recovered or recycled solvents are used in the manufacturing process, their use must be documented.

When customer returns are received, the material must be placed in quarantine to separate them from approved materials. The reason for the return, as well as investigation into the cause for this reason, must be conducted and documented. The material must be evaluated to determine if the quality of material is affected and/or if the material can be returned to stock. Additional testing may be necessary in order to make that determination. Any testing must be documented. If reprocessing or rework is necessary, it must be performed in accordance with the requirements detailed above.

Where materials have been exposed to extraordinary conditions such as extreme temperatures, smoke from a fire, radiation from natural disaster, or other similar incidents, that material should be destroyed. Testing should be designed to be appropriate for the use of the material.

a. Compliance Requirements

All reprocessing and reworking must come to the attention of the quality unit for final review and approval. Continuous reprocessing to bring a batch into conformance should not be allowed, as it indicates that there is something unusual and unknown about the process and/or the quality of the product. The need for additional or new tests must be decided before a reprocessed batch can be released. The quality profiles of reprocessed or reworked batches should be compared to normal, first-time-right batches. This should include purity and impurity profiles, as well as physical profiles. In general, all rework processes require prior regulatory approval.

For each customer return, document the investigation into the reason and cause for the return. Identify corrective actions where appropriate. The quality responsible person should make the decision to return material to stock, further process it, or discard it. While this is not spelled out in the regulations, it should be clear that only the quality unit has the authority to return materials to production, as this serves as a release function.

15. ICH Q7A Section 15: "Complaints and Recalls"

Customer complaints represent a unique opportunity for quality improvement. Complaints can come into a company in a variety of ways: person-to-person, via a telephone call, through e-mail, or regular mail. All quality related complaints received by any employee of the firm must be channeled to the quality department, investigated, and this investigation must be documented. The record must minimally contain the name and the address of the person initiating the complaint, the date the complaint was received, and a description of the complaint. The investigation must include the final decision regarding the material and a copy of the response sent to the complainant. Complaints should be periodically reviewed for trends suggesting areas of improvement.

If the complaint is of a serious nature, which might justify concern of the material on the market, local regulatory agencies must be contacted within time frames stipulated in the local requirements. In rare occurrences, recall of marketed products may be necessary. Each firm must have in place procedures that define how a recall is to be conducted. The person responsible for the recall must be identified in the procedure.

a. Compliance Requirements

All complaints must be investigated and documented by the quality department. The recall procedure is one procedure that a manufacturer does not want to gain experience in implementing. However, when a firm finds itself in a recall situation, decisions must be made quickly and a well-defined procedure can facilitate the communication and decision-making processes. The procedure should identify the sources and types of information necessary for the decision, which functions are to be present in an advisory capacity, and the ultimate decision maker whether to proceed with a recall.

16. ICH Q7A Section 16: "Contract
     Manufacturers (Including Laboratories)"

When another firm manufactures or tests products or materials, the responsibilities identified in this chapter must be defined as falling under the contract giver or the contract acceptor. A formal document typically captures the assignment of these responsibilities and is termed the quality agreement. Each contract manufacturing or testing situation is unique and will require a quality agreement specifically tailored for that situation.

a. Compliance Requirements

The contract giver and the contract acceptor each usually have their own template for their quality agreements. Thus when a contract-manufacturing situation is entered into, the assignment of responsibilities and capturing these into a quality agreement requires negotiation not only for the responsibilities but also the formatting of the agreement itself. There is no right way except to insure that there is clarity from both parties as to the responsibilities.

IV. THE QUALITY CONTROL AND QUALITY ASSURANCE DEPARTMENT

The cGMP regulations define the responsibilities of the quality control and quality assurance department throughout all phases of manufacturing. Part IV of this chapter on quality reviews the specific laboratory controls, and all the QC/QA-related responsibilities throughout the regulations.

A. Laboratory Controls, Taken from the FDA's Regulations for cGMP

The first subsection covers general controls. As expected, all activity associated with the testing of materials must be scientifically thought out. Sampling must be based on statistical grounds, and all procedures should be documented. This includes all activities of the laboratory in its efforts to evaluate all materials, from raw materials to containers, intermediates, in-process controls, and so on, through to the stability testing of the final APIs.

Testing for the release of final products should be performed on each lot produced. Sampling plans, based on statistical grounds, should also include supportive data illustrating that the batch is homogeneous and the process will always yield a uniform grade of material. The test methods must be validated, which means their accuracy, sensitivity, and linearity over a variety of concentrations of material, specificity, and reproducibility have been established. Such criteria apply to all test methods used throughout the manufacturing process, not only to the API.

Stability testing is also required to demonstrate that the material will hold its quality over the labeled storage conditions and time. When establishing the storage conditions for the first time for an API, studies should include the extremes of conditions likely to be seen. The testing protocol should include all and only those attributes that may be affected by the storage conditions. Consideration should also be given to the stability of the product during its planned method of shipment. The regulations offer an adequate amount of flexibility to the storage conditions for the stability study samples. The requirement logically states that the sample container affords the same level of protection, as does the bulk container. The results from these studies are used to determine either an expiry date or a re-evaluation date. Re-evaluation dates are preferred since APIs exceeding their expiry dates are to be discarded, while those with re-evaluation dates may be returned to stock following a satisfactory re-evaluation.

As APIs are tested for purity, impurity testing on a lot-by-lot basis is also required. The expected impurities should be determined for normal production batches, and new impurities should be hunted down when evaluating a process change.

A sample from each lot of API and key intermediate should be taken and stored for annual visual examination, as well as to cover any future investigative purposes (for example, to evaluate a customer complaint on the lot).

Finally, if animals are used for testing purposes (although this is very rare today), they require the same degree of control and suitability for testing as analytical equipment, reagents, and other aspects of the manufacturing process.

1. Compliance Requirements

Laboratory operations must be documented, and the QC leader must approve any changes. Retesting or resampling is a serious matter that can only be conducted following specific conditions and must be carried out under predescribed written procedures. See FDA's Guidance for industry "investigating out of specification test results for pharmaceutical production," issued on their website (fda.gov) September 4, 1998. An effective calibration program for all equipment and reagents is required. The source of reference standards, their storage, and use should be clearly defined. All different physical forms of an API must be included in the routine stability study program. Manufacturing change control systems should evaluate all likely attributes that may be affected, as well as additional attributes that are not part of a typical release protocol (for example polymorphism or new impurities). Initial lots from a process change should be added to the stability study program. An effective program to visually examine each lot of each key intermediate and API must be established, the results recorded, and any cause for investigation taken are also documented.

B. The Quality Unit Responsibilities

1. The Reporting Relationship and General Responsibilities

Throughout the cGMPs, reference is made to "the quality unit." That is the way the FDA and other agencies address the areas of the organization responsible for the quality control, quality assurance, and other QM activities. While the regulations do not mandate to which area or department the quality unit should report, all regulators make it clear that they must have a reporting relationship that allows, and even encourages, independence of judgment. The people assigned the responsibility to judge the quality of a product, material or process cannot be expected to do a good job for the company if there is a conflict of interest between them and their direct supervisors. Regulators and company senior management agree on this point because we are dealing with the manufacture of products used to treat human illness and disease. There is no margin for error.

The overall responsibility of the quality unit is to help the organization develop and implement a solid system of procedures and controls to ensure that each batch of API will routinely meet its predefined quality attributes. To execute that responsibility, the quality unit inherits a broad range of authority: to review and approve all procedures, all systems, all changes, all in-process and final product specifications and test methods, all manufacturing procedures, investigations, and so on. The full scope of authority can only be appreciated by a thorough reading of the FDA's cGMP document. Suffice it to say that all activity related to the manufacture of APIs require the approval of the quality unit.

Clearly, top scientists, technicians, and managers are needed in the quality unit in order to facilitate a smooth-running manufacturing organization.

2. The Quality Control Department

The general laboratories of the quality unit are often referred to as the QC department. This is the area of the quality unit responsible for carrying out the tests on the purchased materials, in-process samples, intermediates, final APIs, and stability studies. In some cases, parts of this testing can be delegated to other departments, such as in-process control tests to the production area. However, the final release decisions are still the responsibility and authority of the quality unit. The quality laboratories are considered part of the manufacturing plant; the QC function is part of the manufacturing process and comes under the same regulations as production areas. They therefore require that systems be described in writing, that cGMP training occurs on an adequate basis, and so on. The test procedures used in this department must be validated. The equipment used must be qualified to demonstrate it functions as designed, and it must be maintained and calibrated on a sufficiently regular basis to assure that it is always working properly. Any problems that occur must be investigated and corrected.

3. The Quality Assurance Department

This part of the quality unit is responsible for the review and approval of the cGMP written procedures and systems used throughout the site for the manufacturing, control, and the release of API's. The QA department typically has responsibility to write quality policy and standards, and to prepare SOPs for the quality control and quality assurance department to follow. The quality unit typically has the responsibility to audit the manufacturing and QC functions to assure that they are following their procedures correctly and they are compliant with other aspects of the regulations, such as performing and documenting investigations where necessary and implementing corrective measures where appropriate.

Another major responsibility of this department is change control management. To effectively evaluate the potential impact of a process change, it is important to contemplate how the predefined quality attributes might be affected—as well as other chemical/physical attributes not normally tested or evaluated. This requires people who have a firm understanding of the chemistry of the process and an appreciation of how a change in the API may affect the drug product process throughout the supply chain.

The significance of the responsibilities of the quality unit and the scope of its influence throughout the manufacturing process, only briefly highlighted in this section, requires that it be staffed with very well-educated, -experienced and -skilled people who are good thinkers, communicators, and confident decision makers.

Other duties of this function usually include the handling and investigation of customer complaints, cGMP training, throughout the manufacturing site and review and approval of major projects such as validation reports or capital investments to assure or improve cGMP compliance.

4. Analytical Technical Service

An analytical technical services department should also exist within the quality unit. The functions of this department include helping manufacturing in troubleshooting to determine "root causes" for quality problems, improving current test methods to make them more efficient or more user friendly, evaluating new technology, evaluating inquiries from official offices such as pharmacopoeia, performing analytical investigations to evaluate complaints against quality, and to keeping all current test procedures up to date.

5. Management of Quality

The head of the quality function has the overall responsibility for the quality systems at the site or across the company. He or she is not the single responsible person to produce a quality product; that responsibility belongs to the head of manufacturing. This distinction may not always be clearly understood. To delineate the division of responsibility more effectively, the head of quality is responsible for ensuring that the requirements are effectively communicated and understood by plant management and an effective compliance and quality system is developed and implemented to achieve those requirements. It is the responsibility of the head of manufacturing to ensure that the system is properly supported, financed, and strictly adhered to. The head of quality, on the other hand, has the sole responsibility to release the product from the plant for further processing, or for sale to a drug product manufacturer. Through the authority to audit, the quality head learns if the systems are being followed; if not, it is his or her responsibility to bring that matter to the attention of the head of manufacturing. If corrective measures are not forthcoming as soon as required based on the severity of the observation, the quality leader is responsible for adequately communicating the matter to a higher level in the organization, regardless of "lines of reporting" as described in organization charts.

A further role of the quality leader is to encourage support and enthusiasm for quality and quality improvement throughout the entire organization. This is best conducted if the quality leader solicits the support and confidence of the manufacturing head, as well as all other technical management at the site and among senior leaderships throughout the organization.

Finally, the quality leader must ensure that his or her staff have access to current information, practices by other companies, and enforcement efforts by the regulatory authorities. This is necessary in order to ensure that the manufacturing system continues to keep pace with the "current" component of the cGMPs.

Appendix A Typical Attributes Evaluated to Control and Assure API Quality

Attribute	How determined	Specification example	Rationale
Description	Visual examination	Agrees with standard or typical product description	A traditional organoleptic "test" with very limited value due to its subjectivity, never-the-less, it remains in use, simply to provide the assurance that the batch appears similar to that expected
Identity	IR, NIR, FTIR, NMR, UV, MS, etc.	Agrees with standard	These electromechanical/ spectroscopic analyses help verify that the correct chemical bonds and arrangement of functional groups are present and elucidate the structure of the molecule
Water content	Karl Fischer	Variable, reported as % by weight, compares to normal ranges	The Karl Fischer test determines how much water is present due to inefficient drying, the hygroscopic nature of the molecule, and/or is chemically bound as a hydrate, but does not distinguish between these types
Residual solvents	GC	Typically reported as "not more than X %"and, if the solvates are controlled/specified in an	Determines how much residual, or volatile, process solvents adhere to the molecule from inefficient

		official pharmacopoeia, their limits would be specified	drying and/or as a bound solvate, but does not distinguish between the two types
Crystallinity/ morphology	X-ray diffraction	Conforms to diffractogram of desired polymorph	The output, called a "diffractogram," provides very precise mechanical information about the shape of the crystal, reporting three dimensions of the crystal's axis and the arrangement/packing of the molecules within the crystal
Melting range	Differential scanning calorimetry	Thermogram and melting curve conform to desired polymorph	The output, called a "thermogram," provides an indication of heat absorption of the molecule during the melting phase; a comparison to a reference standard of the product provides an indication of the sample's purity.
Particle size distribution	Microscopy, laser obscuration, laser light scattering, laser diffraction, etc.	Profile of distribution matches typical results	The output is a distribution of particle sizes, showing the percent present at ever increasing sizes, and a range and mean, reported in microns

(Continued)

Appendix A Typical Attributes Evaluated to Control and Assure API Quality (*Continued*)

Attribute	How determined	Specification example	Rationale
Heavy metal content	USP methods	Typically reported as less than X ppm	Indicates the parts per million of (potentially toxic) heavy metal contaminants/impurities in the drug substance
Palladium (platinum), etc., content	AA, ICP-OES, ICP-MS	Typically reported as less than X ppm	Indicates the parts per million of specifically palladium or platinum (or any specific metal when using the appropriate detector) in the drug substance. A specific metal test is conducted whenever a specific metal is used as a catalyst during the synthesis.
Chiral purity	HPLC or optical rotation	% enantiomeric excess; or degrees of specific rotation, compared to chirally pure reference standard	Provides an indication of the enantiomeric purity of a chiral material, this test is conducted when the drug is designed to exist as a single enantiomer, or when an enantiomerically pure chiral raw material was used in the synthesis, and the chirality or degree of enantiomeric purity is deliberately destroyed to yield a racemic mixture. In this case, the specification would be, for

Impurity	HPLC	Limits based on process averages, safety and/or regulatory expectations; generally reported as total is less than X%, no known impurity greater than 0.5%; no unknown impurity greater than 0.1% (or lower–sometimes more detailed)	example, less than 2% enantiomeric excess. The impurity profile is very important both as an indicator of the purity as well as a show of consistency and reliability of the manufacturing and purification processes
Assay	HPLC	Not less than 98%, not more than 102%, Compared to reference standard of known purity/potency (weight/weight basis, on dried samples)	Proof that the process yields the expected purity
Microbial purity	Viable aerobic counts, yeasts and molds, specific indicator organisms	See USP or EP	Microbial tests reveal the bioburden associated with the drug substance. Biomaterials generally come from process water used in the final step. The test is not necessary if the final step in non-aqueous, and the drying and subsequent handling are controlled such that bio-exposure is limited or not possible

8

Plant Operations

STANLEY H. NUSIM

S. H. Nusim Associates Inc., Aventura, Florida, U.S.A.

I. PLANT ORGANIZATION

The first step is the planning of the organization of the plant. I will, in this discussion, make certain assumptions about the company. Many of the necessary activities required for a pharmachemical business operation, other than the pure production operations themselves, should usually be the responsibility of a separate organization within the company.

There are usually four specific organizational areas required in any plant organization. These are the manufacturing operations themselves, plant maintenance, materials management and quality control and assurance. The last three activities are the subject of individual chapters in this book and I refer the reader to those. I will concentrate on the manufacturing operations and restrict my comments to broad generalities on these allowing the full chapters to cover the appropriate details.

Other key activities, also the subject of separate chapters in this volume, relate to process development, plant design, and regulatory activities that are all critical to the initiation and approval of any plant. However, these activities in any large organization are separate functions that interact with the plant. Process development as well as regulatory operations is traditionally part of a corporate research division, while plant design would generally be part of a corporate engineering function.

The interaction of the plant with the research and development area will be most significant during the design and startup phase of the plant and each of its products. The regulatory activities that would be required for the preparation and filing of the required documentation for the FDA, in the United States, or the Health Ministry of the country where sales are planned will be the subjects of discussion in the appropriate chapter.

II. BATCH VS. CONTINUOUS

The evolution of the industry has seen an interesting shift. The early days of pharmachemical manufacturing, particularly in Europe, were focused on the input of "industrial chemists" who developed the complex multistep processes often required to achieve the unique structure of the product. This, in combination with the relatively small commercial requirements, led to relatively small-scale batch operations with batch equipment in the 100–500-gal size. Generally, the equipment was a relatively small scale up from the laboratory and mirrored the laboratory operation. This was more than adequate to meet the then small market needs. In addition, new product production requiring different equipment needs would be met by building another small factory within the plant and resulted in a plant maintaining many small dedicated operating factories.

As the business of pharmaceuticals grew, particularly in the 1940s and 1950s, the kilogram requirements for the specific pharmachemical leapt. This drove the need for larger vessel volume and more complex batch systems and this drove the capital cost up at a time when equipment costs were already rising.

Furthermore, during these decades other impacts began to weigh on the capital and operating costs of each and every factory in a plant. These were the new issues of environmental impact and control and the increased safety and personnel exposure control mandated for each operating unit. All of these factors increased the cost to operate each separate unit. This gave impetus to allow the growth of the impact of the chemical engineer, rather than the chemist, on the transfer of the synthetic process from the laboratory to the commercial scale. In these years, particularly in the United States, the application of more traditional large-scale continuous manufacturing techniques already in use in the high-volume petrochemical and heavy organic chemical industry was introduced. This became possible only because of much larger requirements for the pharmachemical from thousands to millions of kilograms per annum. In addition, continuous processing was often superior to batch processing because of the increased process efficiencies that were possible and, very significantly, the ability to limit time and temperature conditions to the more temperature-sensitive molecules now often being demanded for the newer products.

Surprisingly, the increased material requirements, driven by explosive product volume growth, have seen a compensating effect in recent years. Although the growth of market volume continues to increase requirements for product, the new research techniques developed in the 1980s and 1990s has resulted in much more potent products greatly reducing the quantities of pharmachemical needed for the market. As an example, the first antihypertensive drug, methyldopa, initially marketed in the 1960s, had a 250–500 mg dose regimen. Hence, for each billion tablets required for the market about 400,000 kg of drug was needed. Today, the most popular medications to control blood pressure are ACE inhibitors, which require only a 5, 10, or 20 mg dose regimen. Thus, for 1 billion 10 mg tablets, only about 10,000 kg of bulk drug is needed.

The net result of this is the refocus on batch operations for most APIs.

III. DEDICATED VS. SHARED MANUFACTURING FACILITIES

A number of factors have led to the shift from dedicated to the sharing of manufacturing facilities for APIs. The first is, again, the increased potency of APIs and hence the reduced quantities required satisfying the market requirements.

The second point is the significant capital cost requirements for a facility. This cost increase starts with normal inflation trends over the years and has added to it a number of other factors. The first is the increased sophistication of the chemistry that accompanies the increasingly complex chemical moieties being synthesized on a commercial scale. Today, what had been strictly laboratory procedures and unit operations such as chromatography has become a routine plant operation.

In addition to these, more nontraditional operations on a commercial scale, are the sharply increased requirements in most countries focusing on cGMP. One then adds as well to the operational issues that of waste management, environmental concerns, and safety and employee exposure.

The final point is the normal economics of larger-scale production. If one continued to utilize dedicated facilities for each product and was even willing to spend the capital needed, they would be faced with the higher unit operating costs associated with smaller batch sizes required to operate for each API that resulted from the reduced requirements of APIs. Utilizing the same facility for multiple products dictates the need for larger batches and the resultant economy of scale.

IV. SHIFT OPERATIONS

The chemical manufacturing industry has long seen the need for round-the-clock multishift operation.

Initially, batch chemical processing saw a need for extending shift operations beyond one shift operation only when the process needs dictated.

However, as the chemical industry and later the pharmachemical industry extended to continuous operations 24 hr round the clock, multiple shifts became common often extending to 7 days.

Later, as the return to batch processing occurred in more recent years, the idea of multishift operation was maintained from an efficiency point of view in order to more effectively use the invested capital in the plant.

There are a number of approaches to scheduling shift work for a 5-day around the clock operation. The simplest, administratively, is to have people assigned to the same shift routinely. This may be efficient to supervise but it forces a fixed day, afternoon, and night "crew". Keeping people on a night-only schedule is difficult to sustain as turnover would be high and would be generally more difficult to maintain from the employee perspective.

This leads to the more common and generally more acceptable three-shift rotating arrangement whereby each week the shift crews are rotated, first from day to afternoon and then, the following week, from afternoon to night shift. This is a more reasonable schedule that has been used routinely throughout plant operations.

The more administratively challenging schedule is to operate on a 24-hr 7-day schedule. Clearly, this is the most efficient of all schedules because it uses all 168 hr in the week without stoppage. The elimination of stoppage itself is an added efficiency as there is always some lost time within the scheduled work program whenever a shutdown occurs. As an example, even in a 5-day 24-hr operation, as the end of the fifth (and final) workday there must be some lost time associated with the facility shutdown for the weekend, as well as time lost on the first day to restart the operation.

The normal "stagger shift" operation I have seen and used involves four "shift crews" who work 8 hr a shift beginning at 8AM (day), 4PM (afternoon), and midnight (night).

Table 1   Seven-Day Stagger Shift

Crew \ Day	1	2	3	4	5	6	7	8	9	10	11	12	13	14
I	D	D	D	D	D	D	D	O	N	N	N	N	N	N
II	A	A	A	A	A	O	O	D	D	D	D	D	D	D
III	N	O	O	O	O	A	A	A	A	A	A	A	O	O
IV	O	N	N	N	N	N	N	N	O	O	O	O	A	A

Crew \ Day	15	16	17	18	19	20	21	22	23	24	25	26	27	28
I	N	O	O	O	O	A	A	A	A	A	A	A	O	O
II	O	N	N	N	N	N	N	N	O	O	O	O	A	A
III	D	D	D	D	D	D	D	O	N	N	N	N	N	N
IV	A	A	A	A	A	O	O	D	D	D	D	D	D	D

D, Day shift; A, afternoon shift; N, night shift; O, off.

They work a full 7 days in succession on each shift with days off between shifts varying from 1 to 4 days. This system, outlined in Table 1 operates on a repeating 28-day cycle. Thus, each employee works 21 days out of 28, working 1 full overtime day per cycle.

This operational system, or variations of it, is commonly used. Its advantage is that it allows for overlapping coverage for sickness, training sessions, and vacation. Thus, if a person is unable to fill his/her shift for any reason the surrounding people each work an additional 4 hr. This permits operation without outside (less experienced) people stepping into an unfamiliar operation. This is important in the manufacture of a regulated drug product, where familiarity and understanding of the operation is critical.

More recently, somewhat driven by the issue of energy conservation that emerged during the fuel crises of the 1970s is a 12-hr shift cycle, one shift starting at 8AM and the second at 8PM. This requires four crews that work 4 days on and 4 days off.

The example is given in Table 2.

This shift arrangement requires much more responsibility as overlapping cannot fill absences, as it would require 18-hr shifts. It requires more flexibility to cover vacancies.

Table 2 Four-Day Stagger Shift

Crew \ Day	1	2	3	4	5	6	7	8	9	10	11	12	13	14	15	16
I	D	D	D	D	O	O	O	O	N	N	N	N	O	O	O	O
II	N	N	N	N	O	O	O	O	D	D	D	D	O	O	O	O
III	O	O	O	O	D	D	D	D	O	O	O	O	N	N	N	N
IV	O	O	O	O	N	N	N	N	O	O	O	O	D	D	D	D

D, Day shift; N, Night shift; O, Off.

However, it is very desirable from the worker's perspective as it provides long breaks every week.

V. STERILE OPERATIONS

In the preface to this book, it is pointed out that fermentation operations will not be dealt with in this volume. This would leave the great bulk of sterile operations out of this text; however, there are some sterile operations that must be considered. However, I would not represent this volume to be an authority on this aspect of pharmachemical processing. More specific references should be pursued by the reader.

The primary focus here will be in synthesized pharmachemical that would go into two types of pharmaceutical products; parental drugs, such as antibiotics or ophthalmic where regulatory authorities require sterile products for topical treatments involving the eye.

In the discussion of sterile operations, one must first recognize that there are two distinctly different reasons for sterile operations. The first is product contamination, particularly if the pharmachemical is used directly as an injectable or intravenous material. Here it is necessary to assure that there are no living organisms present that could pose a threat to the patient. The second is completely different; if, as in fermentation or other biological processing, one is carrying out a biological process one must be certain that no foreign living organism are present during that operation. The efficiency as well as the viability of the organism itself could be affected by the presence of an unplanned biological component. In this volume, we are not addressing the biological processing operations; thus, our focus will be on those synthetic chemicals (as an example antibiotics that will be injected into people).

The approach to achieving sterility in a finished product can be achieved in two different ways. The first is aseptic processing and the second is "terminal" sterilization. Aseptic processing will be discussed here, as terminal sterilization is favored as a more certain method as it eliminates any potential contamination during processing.

A. Terminal Sterilization

The processes used for terminal sterilization are heat, temperatures in excess of 140°C, gamma ray radiation, perchloric acid fumes, and ethylene oxide. These techniques are commonly used for sterilizing metals and plastic tools, medical devices, and instruments. This approach is clearly preferred as it provides a final step that assures sterilization of the product and only demands limited controls prior to that final step. However, unfortunately synthetic organic chemicals, particularly synthetic antibiotics are generally not able to accept these severe conditions without adverse effect on the product. This leads to the need to pursue aseptic processing.

B. Aseptic Processing

The concepts of aseptic processing for pharmachemical processing to assure the sterility of the product is no different than classical parental drug production. A sterile facility has a number of fundamental characteristics; the first is complete separation and isolation from all other operations. This dictates that people, materials, and even the ventilation be independent of all other activities at the site. First, the HVAC system must be totally independent of the main system, it providing only filtered air, generally through HEPA filters that remove essentially all particulate matter larger than 0.21 μm.

Levels of cleanliness have been established as "classes", an industry convention. The "class number" measures the maximum number of particles per cubic meter that may be present. It does not represent the number of organisms.

Particles are more easily and uniformly measurable than organisms and it is assumed that as the number of particles grow the opportunity for some of these either being organisms or having organisms on them increases.

The general standard for a sterile environment in the pharmaceutical industry is Class 100. The general standard for pharmaceutical processing is Class 100,000, while often the preparation area servicing a sterile area would normally be Class 10,000. The purpose of this somewhat tighter standard for a sterile preparation area is to reduce the burden on the Class 100 sterile systems that it services. It should be noted that the electronics industry that deals with microchips is even more concerned with particulate matter (whether or not they are living organisms) and has an even more stringent requirement, Class 10.

Access is restricted to materials that have previously been sterilized by classical sterilization. People gain access only following a complete gowning including feet and head covering. The facility itself must include smooth walls and floors that are easily washable and equipment that must be capable of disassembly for cleaning. It is possible to avoid equipment disassembly by installing sterilizing systems in place of the larger equipment. Each of these sterilize-in-place (SIP) systems must be validated to assure its effectiveness.

There are procedures that must be followed in extreme detail without exceptions in order to assure continuing sterility and limiting people access (the greatest source of organism contamination are people). A facility if used continuously in a sterile condition requires a periodic resterilization. History and validation would dictate the frequency of the need to resterilize. Obviously, if sterility is broken, e.g., by an equipment replacement or even a breakdown of the HVAC system, resterilization would be required before processing could resume.

As can be deduced from the above, a sterile operation is much more costly both in operating expense as well as original capital outlay than a traditional nonsterile pharmachemical facility. Hence, one can quickly see the merit of terminal sterilization that eliminates all of these added burdens for aseptic operation. However, if terminal sterilization is not feasible, then one should work one's way back from the pure product until a point is reached in the process where definitive sterility can be achieved. This will limit the added operational burden and cost required for aseptic processing.

Often this is possible in the very last process step. Most often once the desired pharmachemical is formed, it is then put through a recrystallization step to achieve the high purity level usually demanded. The key is for the crude material to be put into solution and either simply crystallized or carbon treated and then crystallized. A natural opportunity is when the material is dissolved in the final step. The solution can then be passed through a submicron sterile filter into an aseptic environment where new contamination cannot occur. This limits the size and scope of the sterile facility and its associated premium costs.

If such a solution step were not in the process, then one would have to follow the procedure described above to locate the latest point in the process where a sterile filtration can be achieved. Considering that all processing subsequent to that sterile filtration step would have to be carried out under aseptic conditions, it might be more cost effective to add a sterile filtration step at the end of process, then operate a number of steps prior to the end of the process. Clearly, a cost study should be carried out to determine the optimum course of action.

VI. CLEAN ROOM

Today, in the physical plant probably the most significant difference between a traditional chemical facility and a pharmachemical facility is the "clean room" in the pharmachemical plant.

This concept, introduced in the 1970s is derived from the landmark U.S. FDA regulation that established "current good manufacturing practices." This legislation did not in itself define the need for a clean room, but it essentially required it in order to be in compliance.

Fundamentally, the legislation did two things: redefined what is a contaminated product and shifted "quality" from "quality control" to "quality assurance."

Up until the introduction of cGMPs, a "contaminated" product was defined by quality control testing showing that: either the material did not meet the specifications or that there was seen in the testing the presence of a contaminant. This, of necessity, presumes that contamination is uniformly distributed in the product and that the samples tested are representative of the batch.

This "uniform" contamination concept worked in the cases that contamination occurred in the processing itself and was, therefore distributed throughout the product. However, it did not deal with extraneous contamination that could come from the room environment or the equipment that would not be chemical contamination as would easily be picked up by testing. This type of contamination would be random in nature and not subject to normally applied sampling techniques.

Thus, the new legislation defined strict "good manufacturing practices," which provided detailed rules governing the controls that must be in place in the facility, the equipment, the people, the documentation, and the process. It essentially said that if a factory was found to be out of compliance to the GMP requirements, it could not assure that any batch of product was free of contamination; hence, all batches are declared contaminated, regardless of the testing results for any batch.

In terms of facilities, the key issue was the common factory environment. It was the practice of isolating the final API in essentially the same environment as all prior intermediates. This potentially exposed the final API product to the same factory environment, where extraneous contamination ranging from paint chips from equipment to rust from overhead piping and unfiltered solids from earlier process steps could carry forward into the final product.

The most practical way to meet this new requirement was to provide a separate "clean room" within the factory where the environment would be subject to higher levels of control than in the normal operating area.

I had the opportunity of overseeing the establishment of the first "clean room" at Merck's multiproduct plant in

Rahway, New Jersey, in the late 1970s. The facility was tailored after a sterile room concept where access was limited, clean uniforms required with the appropriate hair covers. It had controlled and separated air-handling systems, appropriate wall and floor finishes, no protruding piping or fixtures where dust or dirt could collect, and thorough cleaning procedures.

It was designed to be the finishing room where the final isolation of the API took place and any subsequent finishing steps such as milling, final blending (required for all products), and packaging. Even within this controlled environment, covered containers were used whenever transfer of material was necessary.

The size of the area and the extent of processing that would be carried out in the clean room would be dictated by the process. Logically, the best approach would be passing a solution of the product through a fine filter and taking it through a separating wall into the next vessel that is located in the clean room. Normally, most products have a recrystallization step that would fit the need. In the unusual event that a recrystallization is not part of the process, one would go back into the process to a point where a solution of an intermediate exists. It would then require all subsequent steps to be carried out in the clean room. This could cause a significant increase in cost to erect an expanded clean room. One should consider whether adding a standard recrystallization to the process could be justified as capital avoidance.

Another significant decision is whether to build a dedicated clean room for each product or minimize capital by having a single clean room for all products at the site. This would be generally cost effective for multiproduct factories, particularly if the products have relatively small volumes. Again, the overall economics would govern.

VII. COST CONTROL

The costs of any business operation must be monitored and controlled. A pharmachemical factory is no exception.

The usual costs incurred can be broken into a variety of classifications that I will discuss later.

It is clear that if the plant operates to produce a single product then regardless of the cost system used, the actual all inclusive manufacturing costs incurred will be the basis for the product gross margin (PGM), which is the sales revenues minus the actual manufacturing costs. This is straightforward and does not pose special problems.

In a plant, however, where more than one product is made, one must develop a control to allocate the shared costs to the various products. This is not as simple as it sounds and can influence the perception of the profitability of a product.

The first consideration is to determine the fixed and variable costs involved with every product manufactured. Some items are obvious, others are not. A variable cost is directly proportional to the quantity of a product manufactured such as raw materials and auxiliary chemicals used in the synthesis. For each kilogram of product, a specific amount of each raw material is required. If production of that product ceased, there would be no expense for that raw material. Similarly, the labor used to run the batch is also variable; the labor used is directly proportional to the number of batches run. On the other hand, an essentially fixed cost is independent of the quantity of a product made. This would include supervision at the factory level and clean out chemicals used for turn around of the equipment. However, there are many classes of costs that have both a fixed and a variable component.

The examples in Table 3 list typical cost elements classed as fixed or variable.

These costs are always going to be a factor in establishing the overall cost and expense structure for a factory. One normally prepares a budget for a factory based on the expected volumes for each product. This will dictate the total base activity for the estimating of costs for each product.

I strongly recommend that cost standards be developed for each separate process step for each product ideally at each isolated intermediate (although it is not necessary that each step be isolated). The cost standard will include all of the

Table 3   Fixed or Variable Cost Elements

Item	Fixed	Variable
Raw materials	None	All
Direct labor	None	All
Cleaning chemicals	All	None
Steam	Part	Part
Electricity	Mostly	Some
Water	Some	Mostly
Waste disposal	Mostly	Some
Quality testing	Part	Part
Quality assurance	All	None
Supplies	Mostly	Some
Warehousing	All	None
Dispensing	None	All

above items, where the variable costs will be essentially the same regardless of volume and the portion of the fixed costs that each step will absorb.

The basis for dividing the total fixed costs among products can be done in any way as long as the total factory fixed overhead is fully accounted for. Some common bases are to link the fixed overhead portion to the equipment use time or the direct labor use.

This permits the establishing of cost standards that are useful for production cost control.

The primary variables that should be measured against the "cost standards" as variances are:

- quantity of product (or intermediate) actually made in the step ("yield variance");
- quantity of auxiliary materials used in the step ("charge variance");
- quantity of labor used for the step ("labor use variance");
- unit cost of labor ("labor rate variance");
- the number of batches made in the measured time period compared to the basis on which the standard was established ("volume variance"), which results

in a different absorption of fixed overhead than expected;

- the actual costs of fixed overhead that shared by the products is different from that originally budgeted ("spending variance")

The pharmaceutical plant has an added cost not common to most other chemical factories. This is the need to carry out detailed and extensive cleaning whenever nondedicated facilities are used. These "cleanouts," often including detailed and laborious methods, materials, and specific testing ought to be themselves the subject of separate cost standards.

There are a number of reasons for the establishment and use of cost standards:

- first, to provide a direct and accurate measure of the performance and real cost of each process step in the factory;
- second, to provide management with a tool to measure the effectiveness of an operation and an operating group; and
- lastly, to allow the process development people a powerful tool to predict and measure the value of proposed improvements in the cost of manufacturing of specific products.

This allows management to be able to better judge the value of a process improvement to be sure that it is worth undertaking.

VIII. FIXED OVERHEAD ABSORPTION

In a facility dedicated to a single product, there is no question that product, regardless of product volume, must absorb all the fixed overhead costs. Obviously, as the product volume raises the unit cost of the fixed overhead absorption falls.

In a multiproduct facility this issue is less clear. All of the fixed overhead must still be fully absorbed; however, the mechanism can vary and must be decided upon by management. One such commonly used method is to determine the total labor cost planned for the year in that factory and divide that into the total fixed overhead absorption required for such a factory. Thus, based on the plan for the year each product independent of the others has a specific total overhead burden to absorb (linked by standard labor used). If the quantity produced is less than plan, then the product will not cover the entire fixed overhead it was scheduled to cover and generate an unfavorable variance without adversely affecting other products in the factory.

In the paragraph above, it was shown how in a multiproduct facility the impact of one product not meeting its volume projected in the plan does not adversely affect the other products. However, it can have an effect on the cost of the other products in the following year's plan if the reduced requirements for that product are not made up for by increased volumes of the other products it shared the facility with.

Even if the fixed overhead remained the same for the following year, there would be fewer total units of products made at the facility forcing all of the products to absorb more overhead, thereby raising their apparent costs.

Much of what has been discussed in this section is not unique to pharmachemical operations but a rational approach to quantifying the real cost of a manufacturing activity.

IX. SAFETY

Again, this section deals with an issue largely the same as with any chemical plant. It is not my intent to try to cover in detail the specifics of the subject but to focus on the general issue of safety and point to where differences could occur as a result of the factory making pharmachemicals.

Safety covers a variety of issues but can be separated into two domains: safety as it relates to personnel and safety as it relates to facilities. Broadly speaking, pharmachemical manufacturing because of the economics of the product can afford to deal with exotic materials and chemical process techniques normally too expensive to use in commodity products. This leads to processing techniques such as Grignard, phosgenation even cyanide reactions. These all require handling of extremely hazardous materials which will require both special facilities to assure that the workers are protected from exposure and that the environment is similarly protected with decontamination systems and controlled environment.

A. Facilities Protection

The first issue is the process itself. What fire and/or explosion hazard are inherent in the process? A review of the process will point to obvious issues for fire or explosion such as:

- Hazardous materials:

    1. liquids (acetone, alcohols, etc.) that are flammable and or explosion hazards;
    2. solids such as sodium, sodium hydride, magnesium, acetylenes, etc.;
    3. gases such as phosgene or hydrogen cyanide.

The use of these kinds of materials will require close attention to design to assure explosion protection is provided either by eliminating oxygen by inert blanketing to protect the facilities and perhaps remote operation to protect the personnel.

- Process:

    1. The manufacturing process may allow a step to pass through the explosion limits for the solvent used in a given step.
    2. A process step allows the concentration of an intermediate to an oil in order to eliminate the solvent; however, continuing heating may lead to an explosion potential. This same event could occur unintentionally:

As an example, a common step in organic processing is maintaining a reaction mix at a fixed temperature over a period of time by refluxing the solvent from the batch. This requires the solvent vapors to be condensed efficiently and redirected back to the batch to retain batch volume and concentration. However, if some part of the reflux procedure malfunctions (the cooling medium to the condenser is not available or the return from the condenser to the reactor misdirected or blocked) then the batch will unintentionally concentrate to the point where all of the solvent is gone and a concentrated oil of the reaction mass results. In many cases, this mass has been known to be a potential explosive mixture.

It is judicious, in the development process, when the process is finalized to examine each step for hazard potential of these types and to add safeguards in the design to assure that unintended hazardous conditions are either avoided or planned for in the design.

B. Protection of Employees

1. As described above in the hazardous materials section, designing facilities to handle potentially explosive materials with remote operations or inert blanketing is a key factor in personnel protection.
2. Wearing respirators, air masks, or even full-sealed suits to protect when very noxious materials are used or generated that could result in exposure to the personnel.

Here, one must have sufficient toxicity data to know the nature of the hazard potential to personnel. Planning for the operation includes both preparing for normal operation, which should limit exposure, and for potential deviations in operation that could cause greater exposure. It is essential to be overly cautious when planning for these types of operations.

A fact that has added greatly to this issue is the increased potency that newer APIs have shown. This is partly offset in the pharmaceutical dosage form, as very small amounts of API are required in its pill. However, the active, generated in pure form at the pharmachemical manufacturer presents an additional challenge in processing, sampling, and testing operations at that site. It also adds a challenge to the pharmaceutical finished dosage form manufacturer who must handle and charge this material in a pure form to their process.

X. ENVIRONMENTAL

The environmental issues faced by the pharmachemical manufacturer are no different from that with other fine organic chemical producers. The concerns cover all three possible routes of environmental contamination; liquid waste to sewers, solid waste generated at the site, and air pollution.

In the United States and now elsewhere, environmental impact statements are required to be filed before commercial processes can be utilized. Even after approval, the process operation is subject to being checked periodically by a variety of local, state, and federal review bodies.

In the United States, an environmental impact statement must be filed as part of the original New Drug Application. This must define the controls put in place to handle all three possible routes of contamination.

Usually, liquid waste must meet the local standards for going to municipal waste treatment facility. If it does not meet that standard, the best resolution would be an onsite waste treatment facility. This could either be a separate pretreatment step that destroys the contaminant that makes the waste stream compatible with existing waste treatment systems or definitive destruction using an onsite incinerator.

Attempting to "ship" the liquid waste outside to an approved handler is theoretically possible. However, it raises issues of containment over the road as the waste travels to the outside site. Concerns of leakage or an over the road accident that generates a spill raises all sorts of concerns that will require very careful planning and costly execution. It also assumes that plan is agreed to by all of the various environmental review bodies that must approve such a plan.

Solid waste generated within the processing can be a potential hazard. Again the best solution is decontamination on site or destruction on site with an incinerator that achieves a sufficiently high temperature to assure (proven through testing) that the noxious materials are destroyed.

Shipping solid waste to an outside commercial facility is more acceptable than liquid waste because of the lesser issue of leakage and containment over the road. However, appropriate approvals are also needed.

Potentially contaminated gaseous materials that escape from the process facility must be controlled at source and on site, scrubbing towers through which all vapor effluent from the process are commonly used. The preference would be to destroy the contaminant and convert it to something harmless. Subjecting them to strong acidic or basic conditions or other known materials that are specifically reactive with the contaminant can destroy many organics. Scrubbing is easily set up to achieve this goal.

If the scrubbing system is only capable of capturing the material but not destroying it then the problem shifts to disposing of the scrubbing solution. Then the containment and destruction shifts to the liquid steam generated by the scrubbing system.

9

Materials Management

VICTOR CATALANO

Purchasing Group Inc. (PGI), Nutley, New Jersey, U.S.A.

I. INTRODUCTION

An introduction to materials management can fill an entire textbook. The purpose of this chapter is to introduce the most important aspects of materials management and for each topic to address the unique issues of the pharmaceutical industry. References are provided for those who want more detailed information about materials management in general and specific areas in particular such as materials requirements planning.

Table 1 is a materials management matrix. It shows that a primary objective of marketing is to increase customer service, of manufacturing to decrease operating expenses, and of finance to decrease inventory. It also shows that as marketing tries to increase customer service, there is a tendency for

Table 1   Materials Management Matrix

	Customer service	Operating expenses	Inventory
Marketing	Increase	Increase	Increase
Manufacturing	Decrease	Decrease	Increase
Finance	Decrease	Decrease	Decrease

operating expenses and inventory to increase contrary to the objectives of manufacturing and finance. Likewise as manufacturing tries to decrease operating expenses, there is a tendency for customer service to decrease and inventory to increase. As finance tries to decrease inventory, there is a tendency for customer service to decrease and operating expenses to increase.

Therefore, there is a need to balance conflicting objectives. Materials management requires that conflicting objectives be balanced. The best approach to materials management (and to most aspects of life) is a balanced approach.

II. PRODUCTION PLANNING

The objective of production planning is to coordinate the use of a company's resources (materials, processes, equipment, and labor) to make the right goods at the right quality at the right time at the right (lowest total) cost.

The right quality is key in the pharmaceutical industry. Equipment utilized may be dedicated to a single product or be multipurpose. The multipurpose equipment requires thorough clean out between different products. This clean out and turnaround time between different products can be significant. A balance must be made between longer production runs (fewer clean outs) and generally higher inventory and shorter production runs (more clean outs) and generally lower inventory.

In the pharmaceutical industry, sometimes it makes more sense to have several dedicated lines, rather than one multipurpose line to avoid the cost of clean outs and turnaround.

Production planning is made more challenging in the pharmaceutical industry with the introduction of new products. Trying to forecast the amount and timing for new products is made more difficult because of the need for pharmaceutical products to be approved by the Food and Drug Administration (FDA) before the products are allowed to be sold.

Pharmaceutical companies usually have to provide capacity for the new product before the new product is approved. Sometimes, this initial production capacity may be outsourced to a custom manufacturer. In fact, outsourcing has become more important to the pharmaceutical industry as pharmaceutical companies focus internal resources on developing new products, and take advantage of the manufacturing capacity of external sources for both initial and also long-term production capacity.

III. INVENTORY MANAGEMENT

Nowhere is a balanced approach more important than in inventory management.

I still remember the advice given to me by my manager when I first began purchasing materials for manufacturing. My manager explained that inventories should be kept as low as possible. My manager further explained that if inventories were found to be too high I would probably be given a "slap on the hand." But my manager warned that if inventories should ever drop so low that the manufacturing operation was interrupted or a plant shut down, I would probably lose my job. This advice made it clear to me which way I should err—the manufacturing operation was never shut down because of lack of production materials.

What is unique about the pharmaceutical industry is that the finished product is a product that can literally mean the difference between life and death for a patient in need of a life saving drug. To run out of such a product could be deadly.

Another reason why inventories might be maintained at higher levels in the pharmaceutical industry compared to other industries is that the pharmaceutical industry has historically been more profitable than many other industries. If a sale is lost because of lack of inventory, the opportunity cost is significant. The cost of the finished good is relatively small compared to the selling price of the pharmaceutical product.

It is interesting to note that even the U.S. Government kept inventory of pharmaceutical products, such as morphine, in its strategic warehouses to be prepared in the event of war or other situations where these important products might be needed.

The three primary categories of inventory to be managed are raw materials, work in process (WIP), and finished goods.

When materials are received to be used in the manufacture of a pharmaceutical product, an inspection (and possible testing) is usually necessary before the materials are accepted. Materials may be stored in a separate "quarantined" area until they are released and approved for use. The use of certified suppliers has allowed some materials to be received and accepted without additional inspection or testing. The certified supplier may provide a "certificate of analysis" documenting that the material meets the required specifications. Certified suppliers are usually audited to insure that they meet the requirements for certification of the customer.

IV. PURCHASING/SUPPLY MANAGEMENT

Purchasing/procurement/supply management/supply chain management are some of the names that are used to describe a function that continues to grow in value to organizations in general and pharmaceutical companies in particular.

Purchasing involves obtaining the right material (or services), at the right quality in the right quantities, at the right time, from the right source, at the right price.

Again the right quality is significant to the pharmaceutical industry. The supplier must understand good manufacturing practice (GMP). Supplier selection is critical. Supplier certification is important to many pharmaceutical companies. Suppliers must understand the importance of meeting specifications and controlling their processes to meet specifications consistently. Suppliers must notify the pharmaceutical company of any process changes.

Organizations generally spend a significant portion of their sales dollar in the purchase of materials and supplies, and an even greater share when services and capital are included. A small reduction in purchase cost may have the same impact on profit as a much larger increase in sales. This tremendous potential to increase profits is what has caused many organizations to focus more attention on the importance of purchasing and supply management. Purchasing and supply management professionals are adding value to their organizations through strategic cost reduction.

Some of the cost reduction strategies that have been used include:

    i.   Supplier consolidation: Organizations are reducing the number of suppliers that they do business with. Reducing the number of suppliers for a particular commodity increases leverage. You become a more important customer to the supplier. The supplier can reduce the price based on the increased volume. Dealing with fewer suppliers allows for better supplier management. Purchasing and supply management professionals have limited time, and can only effectively work with a limited number of suppliers. Consolidating suppliers allows more time to be spent with the most important suppliers.

   ii.   Specifications: Specifications are important in cost reduction and management. You should purchase only what you need and not over specify. As an example when purchasing a chemical, the chemical may be offered in several grades with different levels of purity such as 98% or 99%. It is a waste to purchase the more expensive 99% purity chemical if the 98% purity chemical will meet the requirements to produce the desired product.

iii. Standardization: Standardization helps reduce the number of different items purchased. For example in the area of office supplies, instead of buying 10 different pens, purchase only three different pens. This increases the volume of the three selected pens and reduces the price and cost. For process equipment such as pumps, instead of purchasing 10 different pumps, purchase only three different pumps. This increases the volume of the three selected pumps and reduces the price and cost. But this also results in other benefits such as ease of operation and maintenance for operators and mechanics who become more familiar with the three selected pumps. Another benefit is the reduction of spare parts. If you have 10 different models of pumps you will need spare parts for all 10 models. If there are only three different models, you need to hold spares for only the three different models.

iv. Competitive bidding: Organizations can reduce costs through competitive bidding. This requires good specifications and a good scope of work. Purchasing will develop a request for proposal (RFP) that is sent to the preselected and qualified suppliers. Today many of these RFPs are sent over the Internet as electronic request for proposals or eRFPs. There is also an increased use of reverse auctions, where preselected and qualified suppliers will bid down the price of a specific commodity with the business awarded to the qualified supplier with the lowest bid.

v. Negotiation: Negotiation is a key tool used to reduce cost. Good negotiation requires good preparation. Developing a negotiation brief before the negotiation is helpful in preparation for a successful negotiation. The negotiation brief will include information such as the history of the suppliers—current and potential and pricing. It will list the objectives of the negotiation including what must be achieved and what would be nice to have. Negotiation is a skill that can be gained through education (courses such as Karrass) and experience. Successful negotiations are usually positive and are a win–win for both the supplier and the customer. It is usually possible to find a better deal for both parties.

vi. Make vs. buy: Many pharmaceutical companies have chemical manufacturing capability. When a new product is being developed these pharmaceutical companies have the option to manufacture the chemical or to purchase it from an outside chemical supplier. This allows the pharmaceutical company to do the make vs. buy analysis and select the lowest cost alternative. During times when there is much external chemical manufacturing capacity, chemical suppliers are willing to "sharpen their pencils" and offer attractive pricing for chemicals.

vii. Outsourcing: Organizations should decide what business they are in and what is their core competency. This is true for pharmaceutical companies. Pharmaceutical companies may be involved with activities that are not core. As an example, facilities maintenance is an area that is not core that could be considered for outsourcing. To outsource successfully, it is important to develop the scope of the work that is to be outsourced. It is important to include all that is to be outsourced. If you miss something the organization doing the outsourced work will be glad to add what you missed, but usually at a much higher price than if it had been included in the original scope of work. To make sure to include all that you want to outsource, it is important to have an outsourcing team with all the involved parties and especially those who are the experts in what is to be outsourced. The outsourcing team should also include a representative from Human Resources as some employees may be outsourced.

viii. Policy: An organization's policy can have an impact on costs. A good example is a company policy on travel. Many companies have a policy on the class of air travel. A company may not allow and may not reimburse employees to travel by air first class. Business class travel may be allowed only for trips longer than a specified duration such as longer than 8 hr. All other travel should normally be economy. How the policy is set could help save the company significantly, but the policy must balance the needs of the traveling employee with the associated costs. If only economy class air travel is allowed, and employees are required to travel frequently on long flights such as between New York and Singapore, the employees may arrive tired and not be as productive as if they had been allowed to travel on business class.

ix. Long-term agreements: Long-term agreements (3 years or longer) generally result in cost reduction. Suppliers are more willing to offer better pricing and invest more time to help with cost saving ideas. The relationship between supplier and customer becomes more of an alliance relationship.

x. Global sourcing: Globalization has affected all organizations and pharmaceutical companies in particular. Cost savings can be achieved by selecting the best suppliers in the world. Many of the chemicals and other materials that are used in the manufacture of pharmaceuticals are manufactured around the world in such countries as England, France, Germany, Switzerland, Italy, and Japan. In the future, more will be manufactured in emerging markets such as China and India. And the quality from these manufacturers can be excellent. When there are multiple suppliers on more than one continent for a particular chemical or material, purchasing can take advantage of exchange rates to obtain better pricing and reduce total costs.

xi. Reduce freight costs: Reducing freight costs of the incoming raw materials and supplies and the outgoing final products is an important way to reduce overall costs. There are opportunities for savings in all of the various modes of transportation including truckload, less than truckload (LTL), air, rail, ocean, and small package.

xii. Reduce lead time: Reducing lead times is especially important to pharmaceutical companies when introducing new products. This allows organizations to obtain materials needed quickly from suppliers to meet unanticipated demand. Reduced lead times improves the time to market a new product.

xiii. Reduce inventory: There is a need for balance when reducing inventory. Inventory should be kept low to reduce the dollars tied up in the cost of the inventory, and to reduce the carrying costs of the inventory, and to reduce the cost of inventory, which may go bad or expire. But inventory must be kept high enough to keep operations running smoothly without interruption. The inventory of final product should be kept high enough to make sure that every customer that needs the product can get it. For pharmaceutical products this is especially important since these products could mean the difference between life and death for the patient.

xiv. Reduce demand: Reducing demand is one of the best ways to reduce cost. If you are able to negotiate a 15% discount on an item, you can save 15%. But if you can find a way to eliminate the need to purchase that item, you have saved 100%—the full cost of the item. An example is a company that changed its travel policy to eliminate all business class air travel. Prior to the policy change an employee could travel between New York and Singapore on business class. One particular traveler averaged four trips between New York and Singapore a year. After the policy change this traveler saved much more than the difference between the price of the business and economy class fare, because this traveler made only two trips instead of four trips. For the two trips that were not taken the traveler saved the company the full fare.

xv. Reengineer purchasing process: Reengineering can start by documenting or mapping the current process. The next step is to identify and eliminate the nonvalue adding or unnecessary steps. Most employees are being asked to do more with less, so it is important to examine all that is being done, and eliminate the nonvalue adding activities. Another approach to reengineering is to start with a blank piece of paper and list only the value adding activities that should be done.

xvi. Automation/information technology: It is important to reengineer first so that the process that is automated is a good one. If you automate a "bad" process you will just do the "bad" process quicker and consistently. Many companies have installed enterprise-wide information resource systems such as Oracle and SAP. These allow better information to be accessed for better decision making. It allows supply managers to see and better understand the spending—how much of what is being purchased by which suppliers. There are other systems that are being implemented to help direct the purchasing to the preferred suppliers to take advantage of the cost saving agreements that have been negotiated. Many of these transactions are taking place or will take place in the future across the Internet.

xvii. Transfer best practices: To improve performance, it is important to identify best practices and to implement them. If a best practice is found within an organization, it should be applied across the entire organization. Best practices should also be identified in other organizations and should be adopted (steal from the best). There are organizations such as the Institute for Supply Management (ISM) and the Drug Chemical and Allied Trades Association (DCAT) that offer seminars and programs to learn about these best practices. Within ISM there is the Pharmaceutical Forum and the Chemical Group, which along with DCAT offer educational and networking opportunities.

xviii. Energy conservation: Energy conservation is and will continue to be an important way to reduce demand and reduce costs.

xix. Measurements: Measurements are needed to identify the opportunities for savings. Measurements are also used to improve performance internally and externally the performance of suppliers. It is important to measure what you want to manage.

V. DISTRIBUTION/TRANSPORTATION

Distribution involves getting the product to the customer at the right time. Channels are the particular paths in which the goods move through distribution centers, wholesalers, and retailers.

Distribution requirements planning is a system approach that allows for distribution at minimum total cost.

Transportation involves the movement of raw materials from suppliers to production and finished goods to customers.

Transportation involves a variety of modes including air, rail, motor freight, truckload (TL) and less than truckload (LTL), and ocean freight.

Many pharmaceutical products must be maintained within specific temperature ranges. Some of these products must be shipped in refrigerated containers or trailers commonly referred to as "reefers" at a premium price.

With many pharmaceutical companies having facilities in Puerto Rico (because of the tax advantages), ocean shipments between the United States and Puerto Rico are common. Again many of these ocean shipments are made in refrigerated or temperature-controlled containers.

Even though air transportation is usually the most costly form of transportation, some pharmaceutical products are shipped by air for reasons of timing.

The pharmaceutical industry is a global industry. Suppliers are selected from the best suppliers in the world and customers are located worldwide. Therefore, raw materials and final products are shipped all over the world, making transportation an important function within materials management.

VI. INFORMATION TECHNOLOGY

Information technology (IT) is an important area and will continue to become more important as time goes on. Many companies including pharmaceutical companies have been, or are in, the process of implementing enterprise information systems. For a global company to be able to pull together information on a global basis is of great value. For example, to have access to the inventory of a raw material or a finished good quickly on a global basis has allowed companies to reduce the amount of inventory. There is a substitution of information for inventory. The better the information available on inventory the lower the levels of inventory required to satisfy customer requirements. The better the information available on what has been and is to be purchased from suppliers the greater the negotiating leverage with those suppliers.

The particular system selected may not be as important as the need to understand if the system meets the requirements of the company. It is also better if the company implements an enterprise wide system consistently across the entire company.

Some companies have selected information technology products with a "best of breed" approach. They selected the best purchasing system, the best accounts payable system, the best manufacturing resource planning system, etc. In theory, this should yield the best overall system. In fact, several pharmaceutical companies that implemented this "best of breed" approach now realize some of the disadvantages of trying to interface the various systems. Each time anyone of the individual systems is upgraded, there is the difficulty of upgrading the interface between the various systems.

Several of these companies now suggest that the "best of breed" approach may not be the best and that one overall system may be a better approach.

Many pharmaceutical companies are now using the Internet to communicate with and transact business with suppliers and customers. The use of the Internet by pharmaceutical companies will continue to grow.

VII. QUALITY MANAGEMENT

Quality management has become important to almost every industry, but remains even more important to the pharmaceutical industry.

I have attempted to provide examples of the importance of quality in many of the previous sections of this chapter. Quality is so important to the pharmaceutical industry that a complete chapter of this text has been devoted to quality.

Some of the areas to consider with regards to quality include current good manufacturing practices (cGMP), ISO 9000 requirements, auditing, and validation.

Current good manufacturing practices provide the minimum guidelines for the production of drugs that are safe, pure, and effective. The Food and Drug Administration is charged with enforcing all provisions of the Food, Drug, and

Cosmetic Act and regulations. The cGMPs are part of these regulations.

ISO 9000 in an international standard for quality management systems. The standards are not specific for any particular industry, but have been adopted and are used at least by some pharmaceutical companies. ISO exists as a series of standards that covers design and development, production, installation and servicing, and inspection. ISO requires that material identification and traceability be maintained, that suppliers are evaluated on a regular basis, and that training programs are established and documented.

Audits are conducted as a management tool for assessing the quality level of an operation. They are used to identify nonconformance and to make corrective actions as needed, and prevent reoccurrence of potential problems that can adversely affect a product. Audits are conducted internally and externally. Supplier audits may be directed "for cause", such as a customer complaint, for change control, or for a product problem. Audits may be scheduled on a regular basis (e.g., every 3 years) for suppliers of key or critical materials.

REFERENCES

The references are provided for those who desire more detailed information about Materials management in general and specific areas in particular.

1. Adams ND. Warehouse and Distribution Automation Handbook. 1996.

2. Allegri .H. Materials Management Handbook. 1991.

3. Anderson BV. The Art and Science of Computer Assisted Ordering: Methods for Management. 1996.

4. Arnold JRT. Introduction to Materials Management. 1998.

5. Ashley JM. International Purchasing Handbook. 1998.

6. Baker RJ. Policy and Procedures Manual for Purchasing and Materials Control. 1992.

7. Ballou RH. Business Logistics Management. 1991.

8. Bigelow CR. Hazardous Materials Management in Physical Distribution. 1997.

9. Bolten EF. Managing Time and Space in the Modern Warehouse: with Ready-to-Use Forms, Checklists & Documentation. 1997.

10. Bowersox DJ, Closs DJ. Logistical Management: the Integrated Supply Chain Process. McGraw-Hill Series in Marketing. 1996.

11. Burgess WA. Recognition of Health Hazards in Industry: a Review of Materials and Processes. 1995.

12. Carter JR. Purchasing: Continued Improvement Through Integration. Business One Irwin/Apics Library of Integrated Management. 1992.

13. Carter S. Successful Purchasing. Barron's Business Success Series. 1997.

14. Chadwick T. Strategic Supply Management: an Implementation Toolkit. 1996.

15. Clement J. Manufacturing Data Structures: Building Foundations for Excellence With Bills of Materials and Process Information. 1995.

16. Copacino WC. Supply Chain Management: the Basics and Beyond. Apics Series on Resource Management. The St. Lucie Press. 1997.

17. Dobler DW. Purchasing and Supply Management: Text and Cases. McGraw-Hill Series in Management. 1995.

18. Ellram LM, Birou LM (Contributor). Purchasing for Bottom Line Impact: Improving the Organization Through Strategic Procurement. The NAPM Professional Development Series. Vol. 4. 1995.

19. Farrington B (Contributor), Waters DWF. The Services Buyer in the Role of Project and Cost Management. 1998.

20. Fernandez RC. Total Quality in Purchasing & Supplier Management (Total Quality). 1995.

21. Ford WO. Purchasing Management Guide to Selecting Suppliers. 1995.

22. Grieco PL. MRO Purchasing. The Purchasing Excellence Series. 1997.

23. Grieco PL. Power Purchasing: Supply Management in the 21st Century. 1995.

24. Grieco PL. Suppy Management Toolbox: How to Manage Your Suppliers. 1995.

25. Handfield RB. Introduction to Supply Chain Management. 1998.

26. Harmon RL. Reinventing the Warehouse: World Class Distribution Logistics. 1993.

27. Hassab JC. Systems Management: People, Computers, Machines, Materials. 1996.

28. Hickman TK. Global Purchasing: How to Buy Goods and Services in Foreign Markets. Business One Irwin/ Apics Series in Production Management. 1992.

29. Hough HE. Handbook of Buying and Purchasing Management. 1992.

30. Killen KH. Managing Purchasing: Making the Supply Team Work. NAPM Professional Development. Vol. 2. 1995.

31. King DB. Purchasing Manager's Desk Book of Purchasing Law. 1997.

32. Krotseng L. Global Sourcing. The Purchasing Excellence Series. 1997.

33. Lambert DM. Fundamentals of Logistics Management. The Irwin/ McGraw-Hill Series in Marketing. 1997.

34. Lambert DM. Strategic Logistics Management. Irwin Series in Marketing. 1992.

35. Laseter TM. Balanced Sourcing: Cooperation and Competition in Supplier Relationships. 1998.

36. Leenders MR. Value-Driven Purchasing: Managing the Key Steps in the Acquisition Process. The NAPM Professional Development. Vol. 1. 1994.

37. Leenders MR. Purchasing and Materials Management. 1992.

38. Locke D, Locke R. Global Supply Management: a Guide to International Purchasing. NAPM Professional Development Series. 1996.

39. Lunn T. MRP: Integrating Material Requirements Planning and Modern Business. Business One Irwin/ Apics Series in Production Management. 1992.

40. Mulcahy DE. Materials Handling Handbook 1998.

41. Narasimhan SL. Production Planning and Inventory Control. Quantitative Methods and Applied Statistics Series. 1995.

42. Newman RG. Capital Equipment Buying Handbook. 1998.

43. Orlicky J. Orlicky's Material Requirements Planning. 1994.

44. Pilachowski M. Purchasing Performance Measurements: a Roadmap for Excellence. Purchasing Excellence Series. 1996.

45. Poirier CC. Supply Chain Optimization: Building the Stronges Total Business Network. 1996.

46. Pooler VH. Global Purchasing: Reaching for the World. VNR Materials Management/ Logistics Series. 1992.

47. Pooler VH. Purchasing and Supply Management: Creating the Vision. Materials Management/Logistics Series. 1997.

48. Ptak CA. MRP and Beyond: a Toolbox for Integrating People and Systems. 1996.

49. Raedels AR. Value-Focused Supply Management: Getting the Most Out of the Supply Function. The NAPM Professional Development Series. Vol. 3. 1994.

50. Riggs DA, Robbins SL (Contributor). The Executive's Guide to Supply Management Strategies: Building Supply Chain Thinking into All Business Processes. 1998.

51. Robeson JF (Preface), Copacino WC (Editor). The Logistics Handbook. 1994.

52. Romme J (Editor), Hoekstra SJ. Integral Logistic Structures: Developing Customer-Oriented Goods Flow. 1992.

53. Ross DR. Distribution: Planning and Control. Chapman & Hall Materials Management/Logistics Series. 1995.

54. Scheuing EE, Scheuing E. Value-Added Purchasing: Partnering for World-Class Performance. Crisp Management Library. 1998.

55. Steele PT, Court B (Contributor). Profitable Purchasing Strategies: a Manager's Guide for Improving Organizational Competitiveness Through the Skills of Purchasing. 1996.

56. Tersine RJ. Principles of Inventory and Materials Management. 1994.

57. Underhill T. Strategic Alliances: Managing the Supply Chain. 1996.

58. Van Mieghem T, Mieghem TV. Implementing Supplier Partnerships: How to Lower Costs and Improve Service. 1995.

59. Wood DF. International Logistics. Chapman & Hall Materials Management/Logistics. 1994.

60. Woodside G. Hazardous Materials and Hazardous Waste Management: a Technical Guide. 1993.

61. Zenz GJ, Thompson GH (Editor). Purchasing and the Management of Materials. 1993.

10

Plant Maintenance

RAYMOND J. OLIVERSON

HSB Reliability Technologies, Kingwood, Texas, U.S.A.

I. INTRODUCTION

Pharmaceutical plants need to address maintenance and reliability issues better in order to produce life critical products, on time, on cost, and on quality, safely with respect for personnel, property, and the environment. Also, there are issues of evidence of service and other FDA, OSHA, and EPA compliance factors that impact the maintenance function in pharmaceutical plants. These issues are discussed in other chapters within this book.

Our experience in pharmaceuticals shows that plants pursuing manufacturing excellence develop a strategic plan for maintenance and reliability and are focused on reliability-balanced scorecards, maintenance basics, condition monitoring, operator-driven reliability, reliability engineering, and risk management. These issues are basically the same for continuous process plants or batch plants. However, it has been our experience that a sound grasp of maintenance excellence is lacking within the typical pharmaceutical batch type plant. There seems to be a stronger focus on launching new products and meeting production schedules than on day-to-day maintenance.

II. STRATEGIC PLAN

Successful pharmaceutical plants develop a multiyear, strategic plan for maintenance and reliability. The key issues are developed with a cross-section of plant personnel. A proper strategic plan will involve the following issues:

    A.   Vision: Where will we be in 3–5 years?

    B.   Mission: How will the vision be reached?

    C.   Goals and objectives: How will we know when we get there and what it was worth?

    D.   Philosophy: What will our maintenance and reliability culture be?

    E.   Organization structure: How will our structure change during the journey?

    F.   Rewards: How will all personnel be recognized for their efforts and achievements, ideally in a group sense?

    G.   Training: What training will be required to reach our destination?

    H.   Maintenance role: How will the role of the maintenance department change during the next 3–5 years? Will the maintenance department exist as a separate function in 5 years?

    I.   Technology: What role will technology play? An example is interfacing process control computers to a computerized maintenance management system (CMMS) for integrated condition monitoring, which will lead to a sound predictive maintenance program.

J. Capital strategy: What sort of capital spending will be required to reach our goals?

K. Work force strategy: How will hourly personnel be involved? Will we have to negotiate significant change?

L. Customer strategy: How will our customers be involved in our journey? What will our achievements mean to our customers?

M. Vendor strategy: What will change in our relationships with our vendors of services and materials, and how will the benefits be measured?

III. RELIABILITY-BALANCED SCORECARDS

Once a strategic plan is in place, pharmaceutical companies need to implement a reliability-balanced scorecard (RBS) to integrate measures to achieve their vision and mission. The RBS tracks performance and measures results across four distinct areas—financials, business processes, innovations and learning, and customer support. Historically, companies have focused on only one track of measurements. For example, from a shareholder's viewpoint, the focus was always financials. If manufacturing became the focus, then units produced to plan would be a measure. With the RBS, companies have the opportunity to integrate the lagging indicators of financials and customer support with the leading indicators of business processes and innovations and learning.

The RBS provides an integrated set of measurements from unit level to plant level to division level to corporate level. For the first time, the company can tie their budgeting with their strategic plans and track month by month how they are progressing. However, in setting up the scorecard, care must be taken in developing the measures at each level. Pharmaceutical companies need to ensure that their measures meet each of the criteria in Figure 1. Performance measures should relate vision and mission to the strategy and tactics established to meet a set of goals. Performance measures should not all be financial. Each unit or location can have measures that vary from others. They can change over time.

PERFORMANCE MEASURES

✓ They Relate to Strategy
✓ They are Generally Non-Financial
✓ They Can Vary Between Locations
✓ They Change Over Time
✓ They are Simple and Easy to Use
✓ The Provide Fast Feedback
✓ They Should Drive Performance

Figure 1 Performance measures.

For example, if one of the strategies to improve reliability is to implement a preventive maintenance (PM) program, then the first set of measures would be to track the number of equipment items that have PMs established. This would hold until the PM program was in place, then the relevant measure would change to PM completion rates. Next, the measures should be simple and easy to use. This is critical for acceptance at each of the levels that the RBS touches. Performance measures, in all cases, should provide fast feedback. This is necessary for two reasons as follows. As the units review their progress toward established goals, they need to know the impact the change is making. A business process change that can have dramatic impact monthly should not be measured quarterly. The units and management must be able to control the impact of change. Finally, the performance measure should foster improvement. All measures should be used to track and control positive change within the company and the individuals affected (Fig. 2).

By using these principles and building a relevant set of measures, pharmaceutical companies can set the plants on the right path to reliability improvements. For example, effective planning and scheduling of maintenance work can affect several process indicators, such as schedule compliance or percent of planned work. Results indicators include such items as percent overtime or percent "E" work. The financial change that can be tracked through the improvement in planning

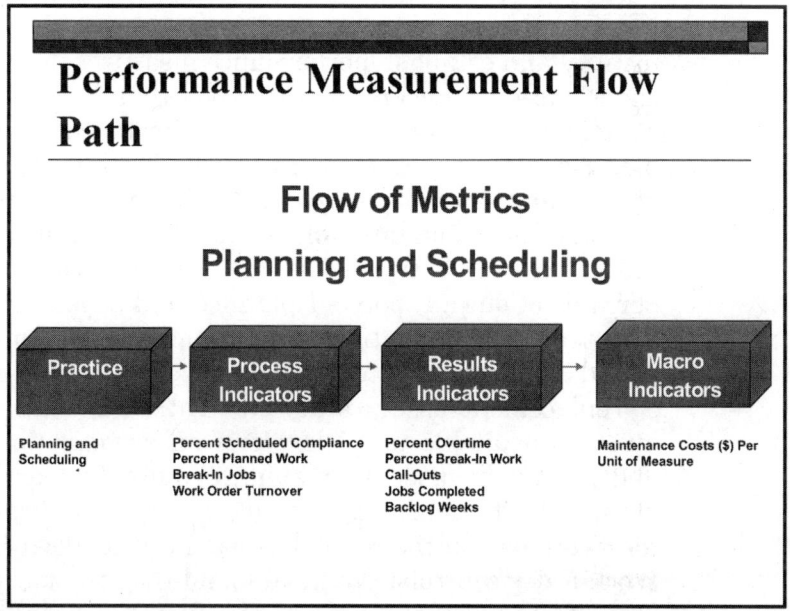

Figure 2 Performance measurement flow path.

and scheduling of work can be measured through percent maintenance cost to replacement asset value.

Through implementation of the RBS, pharmaceutical companies have the opportunity to bring a modern approach of measuring change to their companies. Using the RBS to measure the improvements in maintenance and reliability will provide an integrated view of reliability impact on such areas as financial, customer support, business process, and innovation and learning.

IV. MAINTENANCE BASICS

There are a number of issues that must be addressed in the maintenance basics arena in order for a pharmaceutical plant to achieve excellence in maintenance. They are performance measures, work order controls, preventive maintenance, spare parts management, operations/maintenance relations, training/continuing education, and the CMMS.

A. Performance measures: The first step in establishing an effective maintenance basics program is to identify and establish performance measures. Most experienced managers have learned that "you get what you measure." Successful pharmaceutical plants develop and implement a series of supportive key performance measures to track and manage improvement of maintenance/reliability. Examples are mean time between failure, maintenance cost per unit of output, percent planned and scheduled maintenance, preventive maintenance tasks completed, expense maintenance cost as a percentage of replacement asset value, etc. These indicators must be available to all levels of an organization and should be used to maximize individual and group contributions. It would be best if these indicators were part of the reliability-balanced scorecard.

B. Work order controls: We have found that the most cost-effective plants have a systematized approach to identifying, prioritizing, planning, scheduling, executing, and recording routine maintenance. This would include plant shutdowns.

C. Preventive/predictive maintenance: Cost-effective plants have an equipment criticality ranking scheme and use reliability-centered maintenance (RCM) techniques to determine the equipment that requires preventive maintenance. They use a combination of equipment manufacturers' recommendations, experience, and available PM databases to validate or modify existing PMs and to create new preventive maintenance schedules where necessary; then they faithfully execute their PM programs.

D. Spare parts (stores) management: Getting the right spare part to the right place on time is an important step in effective maintenance materials management. Cost-effective pharmaceutical plants employ proper systems, procedures, and practices relating to the procurement and management of maintenance spare parts. Size of spare parts inventory and other procurement costs are often excessive in pharmaceutical plants because of the ongoing campaigns to produce new products with ever-changing plant equipment. Vigilance is required to manage this situation. Vendor stocking programs (VSPs), consignment inventory, electronic data interchange (EDI), bar coding, cycle counting, and other techniques should be used to reduce inventory levels.

E. Operations/maintenance relations: Solid customer/ supplier relationships, or even better, partnerships between production and maintenance groups are essential in the pursuit of maintenance excellence. The enemy is not inside the plant walls.

F. Training/continuing education: Companies must determine skills training requirements and provide the training. They also must focus on consistent implementation of roles and responsibilities with all levels of the organization to ensure that the right things get done correctly, on time, and safely. A special subset of roles and responsibilities is multiskill maintenance. The pacesetter plants have a flexible work crew with a broad base of skills who are supported by specialists.

G. CMMS: The final factor in maintenance basics, the CMMS, ties all the issues together. Actually, it institutionalizes the behaviors required to achieve maintenance excellence. Typically, it has a work control module, an inventory module, a purchasing module, and ties to financial systems, payables, general ledger, etc. Today, there are several "on-condition" systems such as Ivara's EXP that enhance a CMMS. A key issue with a CMMS or on-condition system is that it can, if used properly, provide the "evidence of service" required by the FDA.

V. CONDITION MONITORING

There are three steps that a pharmaceutical plant should take to implement an effective condition-monitoring program.

1. Identify equipment: The first step in implementing a condition-monitoring program is to determine what equipment will be monitored. It will be critical equipment and generally is about 10% of the total equipment. All manufacturing and environmental control equipment should be evaluated.
2. Select technologies: The next step is to choose the technologies that will be employed. Typically, well-maintained and reliable pharmaceutical plants employ vibration monitoring, thermography, ferrography, particulate count, detailed motor analyses, and environmental monitoring.
3. Implementation: Routes and timing are established, people are trained, and the program is implemented. Readings are taken, analyses are performed, and recommendations for action are provided to maintenance, engineering, and operations. A strict condition-monitoring discipline will lead a plant to a successful predictive maintenance environment where people at all levels of a plant will be able to visualize the resultant equipment "saves."

VI. OPERATOR-DRIVEN RELIABILITY

The items listed under the category of operator-driven reliability are fairly self-evident. The intent is to increase operators' ownership of their equipment and its reliability. The process encompasses many of the activities of total productive maintenance (TPM). Three key areas of focus are:

1. Improving adherence to equipment procedures (start-up, operation, and shutdowns);
2. Improving communication, coordination, and problem solving between production and maintenance;
3. Increasing operators' responsibilities in housekeeping, equipment inspection, and performance of minor maintenance.

VII. RELIABILITY ENGINEERING

The area of reliability engineering finds the pharmaceutical industry well behind the refining/petrochemical industry. A great deal of "catch-up" will be required for the typical pharmaceutical plant. The main areas of catch-up are listed below.

A. Reliability-centered maintenance: RCM provides an engineering, risk-based technique for managing equipment performance. The method takes the large, highly nonintuitive problem of identifying high-risk failure modes and divides it into many small, easily solved problems in order to design a risk-based maintenance plan.

B. Failure mode study: Potential failure modes of critical equipment are ranked according to their risk (probability × consequence). Typically, 80% of equipment's total risk is due to 30% of its failure modes. The failure mode study allows management to allocate scarce resources (labor, materials, and equipment) on a cost-effective basis to attack high-risk potential failures.

C. Analysis of root cause of recurring failures: In most pharmaceutical plants, over one-half of the work orders completed are unnecessary or preventable. Plants need to establish a systematic approach of recognizing/analyzing recurring failures and determining/correcting the root causes. This eliminates unnecessary downtime and reduces maintenance expenditures for labor and materials. Also there is reduced risk of safety and environmental incidents. One way to look at unnecessary or preventable work is to review past work orders and determine whether they were necessary or not. By grouping them into necessary and unnecessary categories, plants are able to determine where to focus on reducing preventable work. Figure 3 illustrates a typical pharmaceutical plant's results typical plant results with necessary work at 51% and unnecessary work cate-

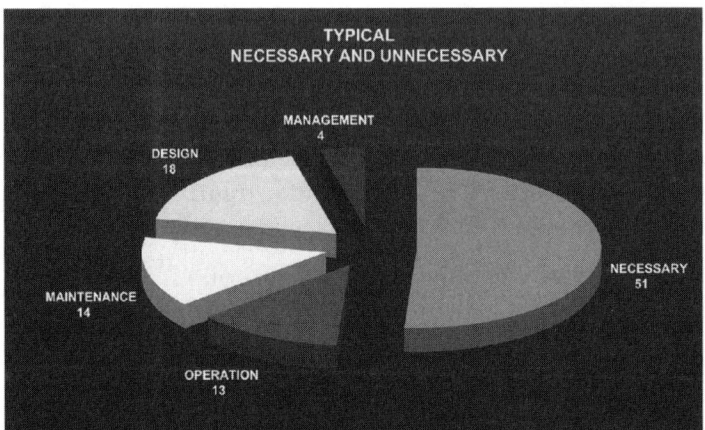

Figure 3   Typical pharmaceutical plant.

gorized as operations caused work (13%), maintenance rework (14%), design problems (18%), and
management lack of support for training, preventive
and/or predictive maintenance, etc. (4%).

D. Maintainability review: This three pronged
approach to improving equipment reliability is
based on failure analysis to identify root causes,
testing of equipment immediately after repair to
ensure quality work was performed, and performance analysis of equipment to determine equipment efficiency rates and replacement intervals.
This methodology uses predictive technologies.

E. Equipment standardization/simplification: A physical assets strategy is developed to include focus
on simplifying and standardizing equipment
throughout a plant. Successful implementation of
the strategy reduces training and repair costs. This
is particularly important in the ongoing "campaign"
approach of the typical batch pharmaceutical plant.

F. Reliability reporting: Special training is offered to
maintenance supervisors and workers to show them
how to measure and track reliability for equipment
in their areas. The focus is on building reliability

indices for key equipment such as pumps, compressors, and motors. Assistance is provided on helping people establish overall equipment effectiveness (OEE) and mean time between repair (MTBR) measures for their areas.

G. Concurrent engineering: New engineering techniques are implemented to ensure that engineering projects are not developed in isolation. Cross-functional teams (engineering, production, maintenance, etc.) are involved in the design, installation, and testing of new equipment to ensure that reliability, maintainability, standardization, performance, and cost specifications for new equipment are met. It is also very important to coordinate engineering efforts at the corporate, plant, and area levels. During a shutdown in a pharmaceutical plant, I observed three different maintenance crews attempting to accomplish three distinctly different modifications to the same unit based on instructions from three separate engineering functions within the company.

H. Bottleneck study: Continuous flow manufacturing (CFM) techniques are used to identify manufacturing bottlenecks, especially those caused by inadequate reliability or maintenance practices. By prioritizing the most critical bottlenecks, appropriate resources can be applied to maximize production throughput.

VIII. RISK MANAGEMENT

The best pharmaceutical plants focus on excellence in maintenance and reliability as a means of achieving manufacturing excellence. This also results in compliance with environmental and safety regulations and preserves a plant's capital investment. Techniques that minimize risks and surprises are:

A. Environmental/safety integration: This is a methodology ensuring that maintenance practices are oriented to satisfying all environmental/safety regulations and requirements. The process involves operations, maintenance, engineering, safety, and environmental personnel in the design of best practices. Maintenance personnel take a greater role in educating operators on how to start up, operate, shutdown, and maintain their equipment more safely. The process addresses EPA and OSHA (PSM) regulations and requirements.

B. Life expectation study: This technical analysis identifies the expected life cycle of critical equipment. Equipment life expectations are developed based on equipment histories, databases, and the manufacturers' information.

C. Life extension engineering: Production, maintenance, and engineering groups work together to devise methods of extending equipment life. Best practices for maintaining equipment reliability and life are developed, implemented, and measured. This methodology is a natural outflow of the life expectation studies.

IX. SUMMARY

Excellence in the maintenance and reliability arenas is best achieved by pursuing a logical, methodical approach, such as we have outlined in this chapter. The pharmaceutical industry must move away from views such as "maintenance is a necessary evil," or "maintenance is an art," to a reasoned empirical approach. The industry cannot afford to squander its heavy investment in research and development by neglecting the maintenance and reliability function.

ACKNOWLEDGMENTS

I wish to acknowledge the contributions of my HSBRT colleagues, Andy Ginder, Keith Burres, Kathy Sorensen, and Shannan Porter in the development of this chapter.

Index